Architectural Interior Systems

Lighting, Air Conditioning, Acoustics

John E. Flynn, AIA

Arthur W. Segil

VAN NOSTRAND REINHOLD COMPANY

NEW YORK   CINCINNATI   TORONTO   LONDON   MELBOURNE

Van Nostrand Reinhold Company Regional Offices:
New York   Cincinnati   Chicago   Millbrae   Dallas

Van Nostrand Reinhold Company International Offices:
London   Toronto   Melbourne

Manufactured in the United States of America

Published by Van Nostrand Reinhold Company
450 West 33rd Street, New York, N.Y. 10001

Published simultaneously in Canada by Van Nostrand Reinhold Ltd.

15   14   13   12   11   10   9   8   7   6

Van Nostrand Reinhold
Environmental Engineering Series

ADVANCED WASTEWATER TREATMENT, by Russell L. Culp and Gordon L. Culp

ARCHITECTURAL INTERIOR SYSTEMS—Lighting, Air Conditioning, Acoustics, John E. Flynn and Arthur W. Segil

THERMAL INSULATION, by John F. Malloy

AIR POLLUTION AND INDUSTRY, edited by Richard D. Ross

INDUSTRIAL WASTE DISPOSAL, edited by Richard D. Ross

MICROBIAL CONTAMINATION CONTROL FACILITIES, by Robert S. Runkle and G. Briggs Phillips

SOUND, NOISE, AND VIBRATION CONTROL, by Lyle F. Yerges

Van Nostrand Reinhold
Environmental Engineering Series

THE VAN NOSTRAND REINHOLD ENVIRONMENTAL ENGINEERING SERIES is dedicated to the presentation of current and vital information relative to the engineering aspects of controlling man's physical environment. Systems and subsystems available to exercise control of both the indoor and outdoor environment continue to become more sophisticated and to involve a number of engineering disciplines. The aim of the series is to provide books which, though often concerned with the life cycle—design, installation, and operation and maintenance—of a specific system or subsystem, are complementary when viewed in their relationship to the total environment.

Books in the Van Nostrand Reinhold Environmental Engineering Series include ones concerned with the engineering of mechanical systems designed (1) to control the environment within structures, including those in which manufacturing processes are carried out, (2) to control the exterior environment through control of waste products expelled by inhabitants of structures and from manufacturing processes. The series will include books on heating, air conditioning and ventilation, control of air and water pollution, control of the acoustic environment, sanitary engineering and waste disposal, illumination, and piping systems for transporting media of all kinds.

Preface

This book is initially concerned with aspects of building design that affect human sensory response and behavior.

There is recognition of the fact that perception and appreciation of a space by an individual or group is dependent, in part, on the interaction of various forms of energy (light, heat, sound); and recognition of the fact that this energy has, as a recent development, become manageable by mechanical means.

One major area of contemporary inquiry, then, is the place of these environmental systems in the various sensory relationships we can classify as *human experiences,* particularly the experiences that we describe as circulation, assembly, recreation, relaxation, meditation, work, and other forms of human participation. Whether this aspect constitutes an architectural problem, an engineering problem, or simply a problem in applied psychology seems to be a matter of semantics. Most certainly, sensory response and behavior have, in the Twentieth Century, become increasingly within the scope of human control; and thus the matter has relevance for anyone who is concerned with the process and technique of building.

But knowledge of this general subject is complicated by the rapid rate of technological and scientific development that quickly makes training, design data, and scientific techniques obsolete. Rather than providing a collection of comprehensive engineering details and data, then, it is the objective of this book to develop an *overview*—a sense of architectural perspective, and a positive attitude for professional judgement.

For this reason, when calculations are shown or implied, they are generally limited in scope and depth to those that are useful for *preliminary* system analysis. It is the intention of the authors to summarize only that information that is necessary to provide a general *sense of the problem.* In this sense, the book is designed to serve as a general reference for architects, interior designers, and consulting specialists who work without an integral interdisciplinary engineering staff. It is also designed to serve as a supplementary text for architectural students.

This book is in total, therefore, a somewhat concentrated interpretative study of the purposes and functions of building in terms of the sensory needs of the occupant and the mechanical technology of our time. In scope, it is intended to *bridge* the technology of environmental control and the art of architecture—to recognize energy as a potential creative medium; to attempt a synthesis of systems and processes that contribute to mastery of the physical environment. It is hoped that such a synthesis will ultimately lead to mechanically-coordinated architectural forms.

Acknowledgments

The authors wish to thank all of those whose interest and help were sources of encouragement during the preparation of this book.

The central thesis of the book was originally developed with the assistance of a 1965 grant from the Arnold Brunner Scholarship Fund and the Architectural League of New York. Subsequent to this, the encouragement and assistance of Luminous Ceilings, Inc. (Chicago) was of great value to the project.

The architectural faculties and students at Kent State University and the University of Michigan provided particularly stimulating forums for the interpretative inquiries that were required in the development of this work. Mr. Samuel M. Mills and Mr. Ernest Kalman further contributed to this development as they assisted in the preparation of diagrams and drawings.

The photographic file of the General Electric Lamp Division was the source of most of the photographic illustrations used.

And finally, the encouragement and active participation of Iris Flynn was an essential factor in the development and preparation of this manuscript.

Contents

PART I

Sensory Performance Standards

Basic Spatial Patterns

Architecture reflects, in part, man's continuing attempt to establish a protected environment that approximates the sensory conditions in which he is most comfortable and at ease (in spite of the fact that these precise conditions appear only occasionally and unpredictably in nature). The environmental control function of any space, then, must be initially oriented toward the occupant. This occupant perceives light as surface brightness and color; he absorbs heat from warmer surfaces and warmer air; and he himself emits heat to cooler surfaces and cooler air. He responds physiologically to humidity, to air motion, to radiation, and air *freshness*. He also responds to sound. And in each case, the intensity of his response depends on his subjective level of adaptation.

A major function of the building is to provide for all of these sensory perceptions, to establish and maintain order in the sensory environment.

In this regard, we can characterize the term *comfort* as implying a reduction of the stresses caused by *negative* influences such as excessive heat, cold, darkness, glare, noise, humidity, etc. The designer must understand the nature of such distracting and disconcerting influences, because over a period of time, they cause strain and fatigue in a participating individual. One objective of the controlled environment, then, is the organization of facilities, forms, and systems to minimize such stresses; for with fatigue, a space or activity can become offensive to an individual, and his emotional attitude toward work or toward an organized activity becomes impaired.

But the elimination of negative influences does not, in itself, insure a pleasant environment. A second objective is even more nebulous, for the designer is in a position to influence the occupant's sense of direction and orientation, and generally to influence the spatial vitality associated with an area or activity. This implies the conscious design of a total building system to establish a *positive* environment where the occupant will tend to

1

feel a sense of well-being conducive to his constructive participation in the activities for which the space is intended.

As we become increasingly aware of the manageable character of the sensory environment, the argument grows that an *appropriate* environment is not limited to conditions where people merely *can* perform a given activity. Rather, it is an environment where people feel encouraged to participate—where the occupant's sensory perceptions and impressions reinforce the behavorial patterns that are inherent in the activity.

These comments are not intended to imply that spatial integrity is achieved through an attempt to create stage-set environments. Rather, they imply recognition and use of the ability to specify and control the characteristics of brightness and shade, sparkle, color, temperature, humidity, air motion, reverberation time, etc. We know that as these influences are made to change, the occupant's awareness of space and activity evolves through his registration of concurrent and successive sensory images.

This first section of the book will focus on these aspects. It will discuss the nature of the sensory background that serves to support human activities and behavioral patterns.

The Luminous Environment

VISION

The sense of vision is based on the eye's ability to selectively absorb and process that portion of the electromagnetic spectrum that we call *light*. This sense is particularly vital, for it is used for most functions that require a grasp of spatial relationships and detail. Initially, this includes the process of orientation and the formation of spatial impressions. Second, there is the process of scanning a variety of information cues, making simultaneous or successive comparisons, and assigning mental priorities regarding importance. Third, there is the process of communication—involving both the identification of meaningful information sources and the subsequent gaining of fine quantitative and qualitative information. And finally, vision is used for interpreting movement and rates of change.

Visual Sensation

The human eye responds to a very narrow portion of the electromagnectic spectrum. The region of 3800 angstroms (deep blue) to 7600 angstroms (deep red) generally defines the *visible spectrum* or *light* energy.

The eye is most responsive in the yellow-green region (5500–5600 angstroms), and the sensitivity diminishes toward the deep blue at one end and deep red at the other (see Figure 1–1.1). The eye is, therefore, essentially unresponsive to both infrared and ultraviolet wavelengths that are immediately adjacent to the visible spectrum. (However, excessive concentrations of infrared energy can heat the cornea and lens, and can cause damage. Similarly, excessive exposure of the eye to ultraviolet wavelengths below 3100 angstroms can cause inflammation.)

Maximum perception of fine detail (visual acuity) occurs when the image falls on the fovea (see Figure 1–1.2). This is the central portion of the retina and is predominantly made up of *cone* cells. Outside of the

3

FIGURE 1–1.1

EYE SENSITIVITY

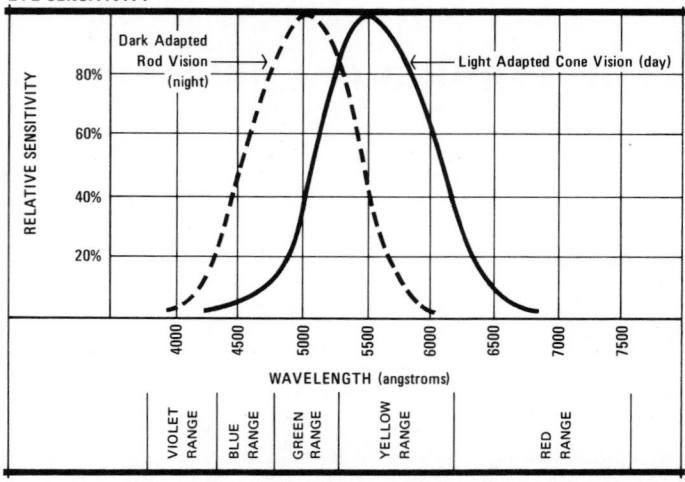

FIGURE 1–1.2

THE HUMAN EYE

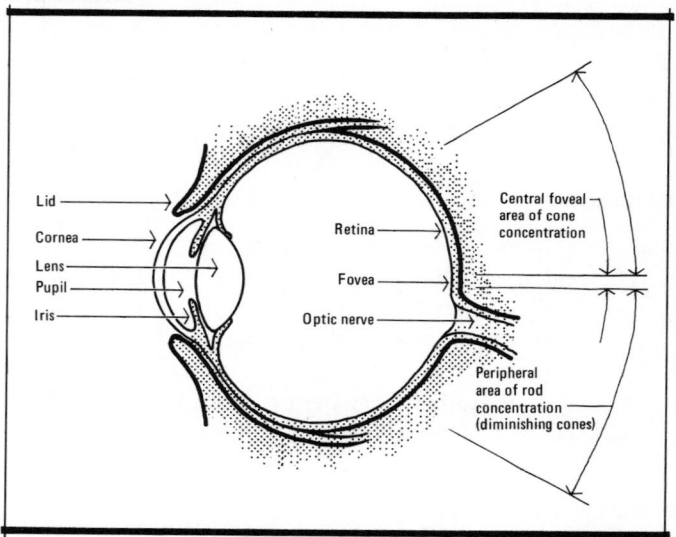

fovea, the concentration of cones diminishes, and normal vision depends primarily on the less responsive *rod* cells.

In general, visual response is dependent on (1) the intensity of radiant energy or flux that is incident on the eye, and (2) the time necessary for the flux to produce a sensation. If exposure time is inadequate, response is diminished (or even negated) accordingly. However, if time is adequate, response depends almost completely on incident flux, which can be translated into surface or object brightness.

Under very low brightness conditions (approaching *night vision* conditions), visual response tends to depend on the rod cells. As brightness on the retina increases above this low-level condition, then, the more perceptive cone cells become increasingly active—and acuity increases sharply until an optimum response condition is approached (see Figure 1–1.30).

Perception of color

In addition to their contribution in the perception of fine detail, cone cells also exhibit good selective response to color. Rod vision, on the other hand, is generally crude vision that is deficient in both detail and color perception.

The areas of cone or rod dominance can be easily observed. Color detail tends to be limited to the central foveal area of human vision, while we perceive little color detail in our peripheral vision where the rods are dominant.

However, as noted previously, dependence on rod vision becomes increasingly dominant at very low intensity levels where the cones lose their ability to respond. In this regard, note in Figure 1–1.1 that peak eye sensitivity for *scotopic,* or rod vision, occurs at a lower 5100 angstroms (dotted line). As a result of this shift in sensitivity, night vision is perceived with a blue or blue-gray cast.

Even under normal brightness intensities when the cones are fully operative, variations in the physiological response to color are noted. For example, colors are not imaged identically on the retina of the eye. Some research studies report that when a green object is imaged clearly, a red object of the same size and in the same position would be imaged slightly larger and behind the retina. At the same time, a blue object would appear slightly smaller and in front of the retina. This optical relationship explains why *warm* colors appear to advance, while *cool* colors seem to recede (and diminish in size).

Response to brightness

The total brightness range to which the eye is sensitive is from faint starlight to sunlight. This is a range of approximately $10^{10}:1$.

However, at any single instant, the brightness range is much more restricted. At any one moment, the eye can readily distinguish a brightness range of approximately 100:1 with good acuity; and if moderate time is allowed for adaptation to either brighter or darker conditions, the eye can accurately distinguish a range of well over 1,000:1.

Adaptation The eye tends to seek a state of equilibrium that is appropriate for the general brightness conditions. This constant adjustment involves some photo-chemical action, but it is most significantly affected by the action of the *iris*, which opens and closes to control the quantity of light that is permitted into the interior of the eye. (The iris opens when the environmental brightness is low, and closes to reduce light penetration when the environmental brightness is high.)

Since there is a time lag involved in this expansion or contraction of the iris, the response of the retina in interpreting brightness at any given moment will be affected by the intensity of brightness in the general visual field during the immediate preceding period of time.

Judgment of brightness differences In the final analysis, then, perception of brightness is a subjective phenomenon that varies with the adaptation of the individual observer. So even under optimum conditions, the eye is a very poor photometer.

For example, when the human eye is called upon to estimate an equal step in brightness, it will seriously underestimate the difference. Normally, given a white surface (80% reflectance) and a black surface, and then asked to select a gray finish that is approximately midway between these two, many observers will tend to select a 15–20% reflectance gray. These individuals may then be quite surprised at the apparent lightness of the 40–45% gray tint that is the true photometric midpoint.

The ability to perceive minor brightness differences is also somewhat erratic and variable. When the occupant is visually adapted to high intensity daylight-type conditions, it may be possible for him to perceive detailed brightness differences that vary from the average by as little as 1%. When lower environmental brightness conditions are involved, however, minimum perceptible differences must vary 5–10% (or more) from the average.

Glare Visual response to brightness also depends on the distribution of flux over the retina. When extremely unequal brightnesses are present in the visual field at the same time, the more extreme excitation of one part of the retina may inhibit the performance of other areas. When this occurs, perception of lower intensity detail is seriously impeded by the brighter *glare source*.

This condition is typically observed by drivers who are confronted with oncoming bright headlights on a darkened highway. The resultant overstimulation of a small portion of the retina makes perception of environmental detail extremely difficult.

Effect of aging The physiological capabilities of the eye tend to deteriorate with age. As a result, there is a reduction in speed of perception, a reduced resistance to glare, and a lengthening in the time required for adaptation. Because of these factors, there is a measurable reduction in visual sensitivity (see Table 1–1.1).

This deterioration is particularly evident for night vision and in very low brightness environments.

Table 1-1.1 Typical Effects of Aging (Visual Sensitivity)

Approximate Age	Relative Visual Sensitivity to Detail
20 years	100%
30	95%
40	90%
50	85%
60	75%
70	60%
80	50%

The visual field

Normal human binocular vision involves a field of view approximately defined as 60° upward from the line of sight, 70° downward, and 180° horizontally (see Figure 1–1.3). However, acute perception of fine detail (foveal vision) takes place within an extremely small angle that parallels the line of sight. This angle is not greater than 2° in diameter, and is generally so small that the eye must change position in order to focus successively on both dots of a colon (:) viewed at a distance of 14 inches.

Outside of this small area of precise perception, then, is a very much broader area of peripheral vision in which perception of detail and color becomes successively less certain. The more central portion of this peripheral vision is defined as a cone of approximately 30° above, below, and to each side of center. This cone (the *near surround*) involves relatively clear imagery. A surrounding cone of general comprehension extends approximately 60° above and to each side, and 70° down from the line of sight. Beyond this, an area extending approximately 85–90° to either side of each eye registers major forms as indistinct masses.

FIGURE 1–1.3

THE VISUAL FIELD

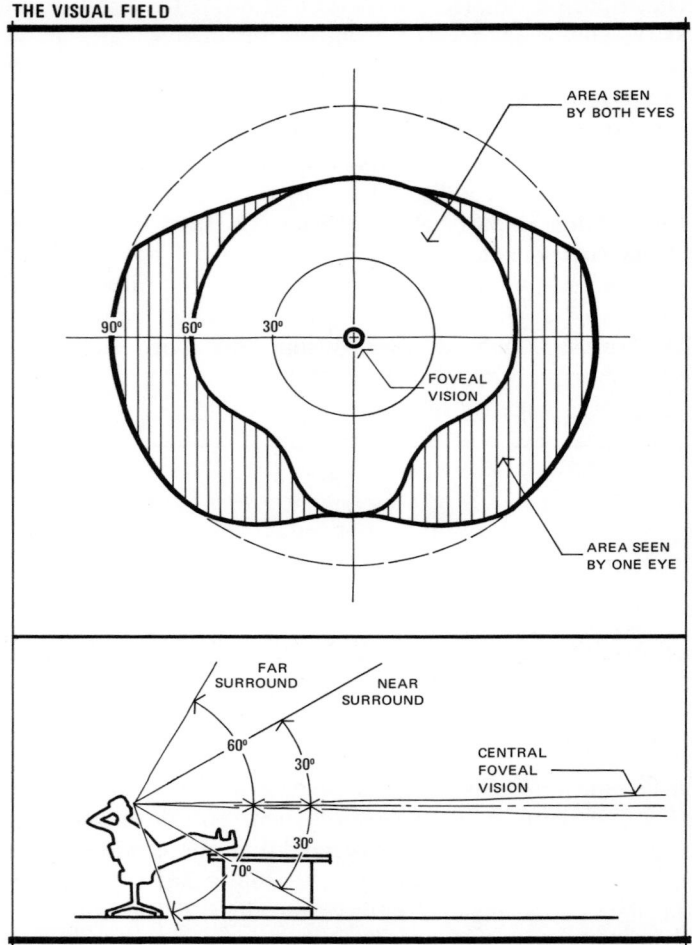

Within the more peripheral portions of the field of view, the eye identifies information cues by responding to changes in brightness pattern and intensity (which makes this area relatively sensitive to movement, flicker, etc.). Peripheral vision is therefore most useful as an influence on the occupant's ability to maintain (1) a sense of general orientation, and (2) a sense of relationship to the dynamic activities in the space.

Perception of fine detail in the visual field   In moderate and higher brightness environments (i.e., conditions where cone vision is predominant), maximum acuity occurs for detail that is located in the direct line of sight (0° from the fovea). Acuity diminishes as the detail is moved to the periphery—and at about 15° from the direct line of sight, acuity is approximately 15% of maximum.

Under very low brightness conditions (i.e., conditions in which rod vision becomes dominant), visual acuity is about 10% of normal maximum, and acuity peaks for detail located about 4° from the fovea.

Detection of movement While peripheral vision is highly useful for detecting movement or similar changes in the general background, perception of such changes remains most sensitive near the direct line of sight. Movement rates of approximately 0.9 minutes of arc per second can be detected in this central area, while movement must approximate 18 minutes of arc per second to be detectable in the more peripheral areas.

A similar pattern affects visual perception of detail in motion. Acuity of foveal vision for an object moving at a rate of 50° of arc per second is approximately 60% of that achieved for a stationary object. When the moving object is located in the periphery, reductions in acuity of this magnitude occur for objects moving at much slower rates of speed.

The Luminous Environment

In general, then, human vision involves a narrow area of sharp central vision (foveal vision) and a much larger, out-of-focus background (peripheral vision). Foveal vision includes the sensation of detail and color (cone vision), while the more inclusive peripheral vision (rod vision) is essentially responsive to brightness and mass. Within this context, background brightness patterns tend to become initially significant because they affect the general sense of spatial orientation which guides the occupant.

To develop this further, background brightness patterns are important because foveal vision is guided or *cued* by information gained through peripheral vision. Potentially significant visual patterns are initially identified and located by scanning and assimilating the total visual field. Then central foveal vision is focused on the relevant detail that has been identified in the periphery. The ensuing study (using foveal vision) is the means through which the occupant derives most of the information necessary for specific orientation and for performance of a specific activity or task. But even while this detailed study is under way in the foveal area, peripheral vision is continually used to maintain general orientation and to identify any new information cues in the environment.

Design of the luminous environment is therefore primarily concerned with two aspects of human sensory behavior:

(1) The visual task of *spatial orientation,* which requires the designer to be concerned with the effect of light in defining the space, the structural enclosure, or the activity (without introducing irrelevant patterns or visual confusion).

(2) Detailed *central task vision,* which requires the designer to be concerned with the effect of light in defining significant information centers and in assisting the accurate communication of visual detail required for acceptable performance of normal activities.

The balanced manipulation of these visual conditions should provide for the occupant's need to judge distances and recognize relevant objects, materials, colors, and forms. At the same time, this environmental balance should reflect the need to protect the occupant from glare and from meaningless visual cues that may confuse his sense of orientation and purpose.

PERCEPTION OF THE LUMINOUS BACKGROUND

An occupant interprets the general environmental background largely through the dominant brightness relationships. The subjective sensation of visual space, then, is primarily a function of brightness pattern and pattern organization—the relationship of surfaces lighted or left in relative darkness.

Vector Influences and Design Identification of Information Cues

Brightness contrast in lighting design can establish a sense of visual direction and focus in a space. This can be complementary and desirable for highly organized activities that involve fixed focal centers and a high degree of visual concentration on the part of the occupant. Figure 1–1.4 shows an environmental setting where high contrast focal emphasis tends to dominate the observer.

On the other hand, lighting systems and surface finishes that reduce or eliminate contrast (by increasing the general brightness level and reducing shadow areas) produce a diffuse or low contrast environment (see Figure 1–1.5). Since all areas tend to be somewhat equal in visual emphasis, such a space encourages more casual or loosely controlled circulation and free selection of points of interest on the part of the individual.

Effect of brightness contrast

The eye is involuntarily drawn to bright objects or to areas that contrast with the general background. Use of this technique can be effective in situations where it is desirable to direct the observer's attention and interest to predetermined detail, while subordinating other items, areas, or surfaces. Table 1–1.2 indicates the relative significance of various ratios of adjacent spatial contrast.

FIGURE 1–1.4

THE HIGH CONTRAST ENVIRONMENT

FIGURE 1–1.5

THE LOW CONTRAST ENVIRONMENT

Table 1-1.2   Brightness Ratios (Vector Influence)

	Vector to Background Brightness Ratio
Barely significant contrast (Low contrast environment; see figure 1-1.5)	2:1
Minimum meaningful contrast in the sense of spatial vector influence	10:1
Dominating contrast (High contrast environment; see figure 1-1.4)	Approaching 100:1

Effect of color accents

A vector influence can be produced with color, as well as with brightness contrast. In this sense, colors of light exhibit varying attraction values that are completely independent of brightness (see Table 1–1.3).

Although the results of such studies are somewhat variable, they indicate that saturated colors of red, green, or blue light will compete for attention with *white* light of greater intensity. Yellow light must be slightly brighter than white and considerably brighter than the other colors for equal attraction value.

Table 1-1.3 Color (Vector Influence)

Color of Light	Relative Brightness Required for Equal Attraction
White light	1.0 ft–lambert
Yellow light	1.2 ft–lambert
Red light	0.3 ft–lambert
Green light	0.4 ft–lambert
Blue light	0.6 ft–lambert

Spatial Order and Form

If brightness contrast is somewhat decisive in establishing the relative visual priority of various information cues, then the organization of brightness patterns becomes a fundamental consideration in defining visual space.

An individual defines his environment through a process of additive perception. He derives successive items of information by scanning the boundaries of a space or activity—thus assembling a conception of direction and limits.

In this sense, the lighting system should be designed to adequately establish the physical boundaries of the space and to identify meaningful information cues. This assists the occupant in maintaining a sense of direction and essential spatial form, and he can participate in the activity with a minimum of visual interference or distraction from the environment.

Visual clutter

The lighting design should generally be such that meaningless or confusing spatial cues are minimized.

Clutter in the visual field is analogous to noise in a sonic evaluation. This is a particularly significant factor when intensive visual effort is required. Above certain levels of information input (which vary with the individual's concentration at a given point in time), visual performance appears to decrease in proportion to the increase in random visual cues that the mind must assimilate and process (see Figure 1–1.6). These reductions in visual performance often reflect an increase in search time. For this reason, when intensive vigilance or complex tasks are involved (such as those involving safety, high performance in spatial comprehension, or complex work tasks), the visual field should be simplified by minimizing irrelevant or meaningless cues. Spotlighting, luminaire patterns, wall lighting, etc. should, in these situations, be developed to simplify the process of orientation and spatial definition.

FIGURE 1–1.6

PROCESSING OF INFORMATION CUES

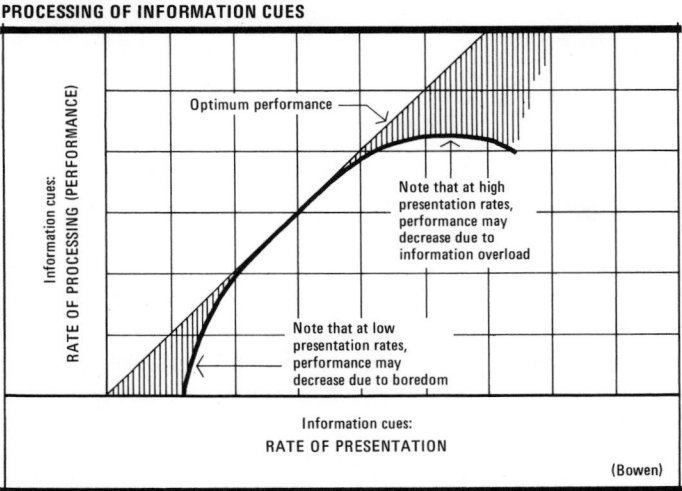

Exceptions occur when more casual activities are involved. In these cases, novelty effects may become useful as a stimulant to relieve spatial monotony and boredom.

Figure 1–1.7   Is an excellent exterior example of visual clutter created by irrelevant and chaotic brightness patterns. In Figure 1–1.8, the more organized and meaningful placement of luminaires tends to provide a better sense of spatial orientation and direction.

As a general finding, then, the ability to effectively perform visual tasks appears to be less consistent in a poorly organized visual space. And related to this is research evidence that individuals tend to voluntarily spend significantly less time in a visually disorienting or cluttered space.

FIGURE 1–1.7
VISUAL CLUTTER

FIGURE 1–1.8
IMPROVED SPATIAL ORDER

Lighting systems as a factor in spatial definition

In developing a sense of visual order, it is important to initially separate the illumination function of the lighting system from the problem of luminaire placement. While fixtures are often placed overhead (on or in the ceiling), very often the objects with which we associate for visual orientation are located in the lower part of the visual field.

At the same time, the eye is drawn to bright areas (such as luminaires) that contrast with the general background condition. Accordingly, if the luminaires are to be a relevant part of the visual background, the brightness pattern must merge with and reinforce the spatial patterns that are meaningful in the activity.

Luminaire patterns   Since the lighted luminaires in the room may become a significant determinant of visual form, the layout of these luminaires should establish a sense of scale, visual order, and appropriate direction (if any).

For example, an activity-oriented location of luminaires may contribute to a sense of spatial organization when the focal points remain stable (see Figure 1–1.9). However, when activities are subject to change and the space must be versatile, moderate to low brightness luminaire patterns may merge with the ceiling plane in such a way that they become a unified part of this major architectural form (see Figure 1–1.10).

FIGURE 1–1.9
ACTIVITY-ORIENTED LOCATION OF LUMINAIRES

FIGURE 1–1.10
REPETITIVE GENERAL LUMINAIRE PATTERNS

FIGURE 1–1.11
REINFORCEMENT OF SPATIAL PERSPECTIVE

When the activity involves circulation or movement, the layout of luminaires can reinforce the sense of direction and spatial perspective (see
Figure 1–1.11).

Lighted surfaces   There is also the action of the lighting system in
delineating the form and character of major surfaces that define the space.
In this respect, the lighting should help to define and separate major
surfaces (see Figure 1–1.12).

Related to this, the *form* of the light distribution (i.e., a *cone* of light,
a linear *wedge* or *wash* of light, etc.) should generally relate to the form of
the affected surface. A wall or ceiling surface, for example, should usually
be lighted with a linear wash of light that approximates, as near as possible, the form and dimension of the surface involved. *Scallops* and similar
irrelevant variations should be minimized; for, except when special effects
are involved, the surface should be basically perceived as an integrated
form, not as a form or surface intersected by patterns of light.

This is not to infer that an irregular pattern of light is never desirable
for its own sake. The shaft of sunlight may have intrinsic value of its
own, as may some man-made patterns that have no relationship to the
physical form of the space. But the real value of such influences is usually

FIGURE 1–1.12
THE INTEGRITY OF THE LIGHTED SURFACE

FIGURE 1–1.13
DEFINING THE JUNCTURE OF MAJOR SURFACES

a novel effect that serves as a temporary spatial stimulant. More lasting and meaningful irregular spatial patterns should, on the other hand, relate to some relevant characteristic of the physical space, such as an interplay with a framed picture, sculpture, or planting—or articulation of the juncture of major architectural surfaces (see Figure 1–1.13).

Spatial distribution of brightness

The development of information cont 't in the visual field may vary with the behavioral needs of the activity. i this sense, the background can be developed to either emphasize or su: ordinate various aspects of the environment or activity.

As one example, a system can be des gned to produce horizontal illumination over the normal work or circulation plane, while subordinating the vertical and overhead elements (see Figure 1–1.14A). The more intensively illuminated central area causes people and activities in the lower visual field to become the dominant brightness feature. The enclosure itself appears as a neutral or even subordinate visual influence. This situation will tend to increase the occupant's normal awareness of nearby detail,

FIGURE 1–1.14A
BRIGHTNESS EMPHASIS: HORIZONTAL PLA

FIGURE 1–1.14B

BRIGHTNESS EMPHASIS: VERTICAL PLANE

of other people, and of their movement. In a relative sense, this condition seems to encourage an attitude of gregarious involvement among the occupants.

The configuration of the somewhat impersonal architectural space itself is, on the other hand, interpreted through vertical and overhead brightness relationships (see Figure 1–1.14B). When visual emphasis is focused on these areas by lighting the major peripheral surfaces (while reducing the

Table 1-1.4 Brightness Ratios (Typical Behavioral Patterns)

	Horizontal to Background Vertical Brightness Ratios
Tendency toward a gregarious behavioral pattern and active movement	1:1 to 100:1 Horizontal is brighter
Tendency toward a sense of detached privacy, introspective behavioral patterns and more relaxed movement	1:20 to 1:100 Background is brighter

intensity of horizontal illumination), foreground detail is lost as objects and people in the central area tend to go into silhouette. This condition therefore causes the foreground activities to become visually subordinate to the general environment or space. Evaluating this again in a relative sense, this condition tends to induce a more introspective attitude in the occupant—making this a more intimate atmosphere in which the individual feels a sense of relative privacy or anonymity.

General intensity of spatial brightness

The subjective sensation of space is also affected by the intensity of brightness patterns, as well as the previously discussed organization of lighted areas and surfaces. Although the basic distribution characteristics of the lighting system may remain constant, significant variations in the general level of brightness will affect the intensity of interreflections in the space. This interreflected light is a diffusing influence and alters the impression of brightness contrast.

At the low end of the brightness scale, for example, the luminous *glow* induced by a low dimmer setting reduces the diffusion inherent in a multi-directional lighting system and creates a high degree of contrast in a room (see Figure 1–1.15A). Higher direct intensities, on the other hand, increase the intensity of interreflections and tend to reduce shadow and silhouette (see Figure 1–1.15B).

Furthermore, at the low end of the general brightness scale, a slight increase in general intensity will produce a vast improvement in the individual's ability to discriminate detail and color. As brightness increases, the *rate* of improvement diminishes, and the environment approaches a condition of maximum acuity regarding the spatial background.

As a result of this combination of behavioral influences, high general intensities will tend to contribute to a sense of increased activity and *efficiency,* while lower general intensities tend to reinforce an attitude of slower-paced activity.

Color Atmosphere

Subtle changes in the color tone of light (i.e., *whiteness*) can influence the subconscious judgment of the general environment. Perceptual awareness of this aspect of light is most intense when a change first occurs or when the individual first enters the space—before the eye has time to adapt to the new condition. (See Chapter 1–4 for a discussion of transitional vision.)

In a perceptual sense, these variations in color of light may recall the

FIGURE 1–1.15A
LOW SPATIAL BRIGHTNESS
INTENSITY

FIGURE 1–1.15B
HIGH SPATIAL BRIGHTNESS
INTENSITY

pinks and purples of a sunrise or sunset; they may suggest the impressions associated with *warm* sunlight or *cool* overcast skies; or they may produce a completely unnatural visual situation.

Each of these conditions can be produced within the broad category of *white* light, although each is actually deficient in some portion of the spectrum. This shift in spectral emphasis is most immediately noticeable in the perception of neutral surfaces and familiar surface tones (such as human skin tones). In fact, it affects the perception of all surface colors and color differences in the space, *graying* some colors while increasing the relative vividness of others. (Also see subsequent discussion: "Perception of Color and Color Contrast".)

These subtle shifts in the perception of surface tones and colors will, in turn, affect the sense of *warmth* or *coolness* that we associate with the visual space. We tend to associate a *warm* visual atmosphere with hues of yellow, through orange and red, to red-purple. *Warm* light sources (like the sun or the incandescent lamp) tend to create a dominant impression of visual warmth by emphasizing these hues while graying others. On the other hand, *cool* light sources (like skylight and some fluorescent lamps) emphasize the colors that tend to create a cool visual atmosphere—from hues of blue-purple, through blue and blue-green, to yellow-green.

Color of light as an influence on spatial interpretations

Although investigation of perceptual relationships are still inconclusive, there are general indications of a number of broad behavioral patterns or tendencies associated with variations in color of light. These are summarized in Table 1–1.5.

Table 1-1.5 Color WHITENESS (Typical Behavioral Patterns)

	Cool-Toned Light	*Warm-Toned Light*
Estimation of environmental temperature	Tendency to under-estimate	Tendency to over-estimate
Estimation of back-ground noise intensity	Tendency to under-estimate	Tendency to over-estimate
Estimation of space size	Tendency to over-estimate	Tendency to under-estimate

Color of light can also affect the subjective interpretation of brightness intensity. Experiments indicate that for each color temperature (i.e., source *whiteness*), there is a maximum and a minimum agreeable level of illumination or general brightness. If the level is too high, surface colors will seem faded and unnatural. If it is too low, the space will appear dim or cold.

Figure 1–1.16 indicates the general character and range of this variable quality of light. Primarily, it suggests that *warmer* light is more acceptable when the brightness level is low.

It further suggests that a room lighted to a uniform horizontal intensity level of about 20 ft-c would be unpleasant either with kerosene lamps

FIGURE 1–1.16

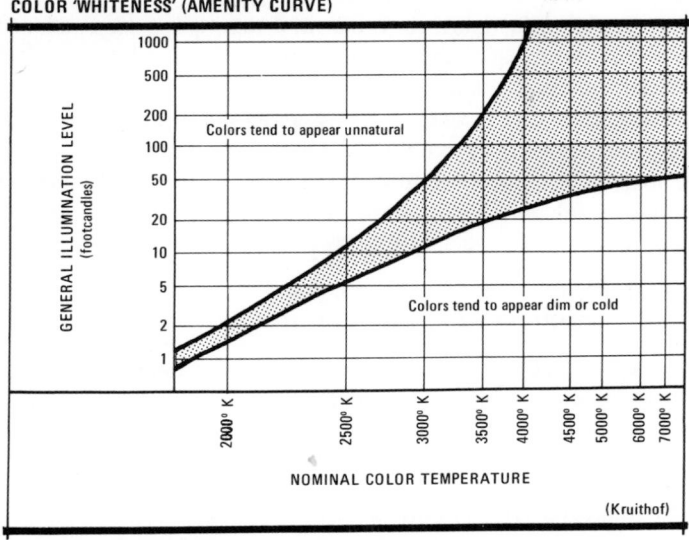

COLOR 'WHITENESS' (AMENITY CURVE)

(about 2000°K) or with lamps that simulate daylight (about 4500°K). When the warm-toned kerosene source is used, the level would seem too high and the space too bright. The same space with simulated daylight, on the other hand, would seem dark and dingy. *Warm* fluorescent (about 3500°K) or general-line incandescent lamps (about 3000°K) would fall within the 2700–3700°K color range indicated as acceptable in the diagram.

Excessive Brightness and Glare

While brightness and brightness contrast are basic in visual communication, excessive contrast or excessive background brightness can disrupt the ability of the eye to perceive fine detail. In the extreme, these *glare* conditions can temporarily cripple vision by destroying the observer's ability to adequately perceive a task, an obstruction, an object, or a space.

The experience of driving toward a late afternoon sun, the after-dark effect of approaching bright headlights, or the occasionally overpowering brightness caused by the sun on clean sand or on white snow are three common experiences that reveal the nature of *disability glare*. This is glare of sufficient intensity to impair visual acuity and to impair the occupant's ability to orient himself.

A somewhat more subtle form of glare is experienced with unshielded fluorescent lamps or excessive luminaire brightness. Although the resulting disability tends to be relatively moderate, temporary physical disability again results due to unequal excitation of the retina (see previous discussion of "Glare" under "Visual sensation").

Glare is generally corrected: (1) by reducing the source luminance (such as by dimming); (2) by the use of baffles, louvers, or diffusers to reduce excessive luminaire brightness; (3) by source relocation outside of the normal visual field; and (4) by reducing the reflectance characteristics of excessively bright surfaces.

Brightness tolerance as a function of area

In estimating and evaluating brightness tolerance, there is a fundamental relationship between *brightness intensity* and *area of brightness*. This relationship will affect the actual quantitative limits of visual comfort.

Table 1-1.6 Limits of Visual Comfort (Area)

Diffuser Material	*Lamp Density (Luminous Ceiling) Maximum Allowable Generated Lumens/sq ft of Diffuser Surface*
White diffusing plastic, flat	600 lumens/sq ft
White diffusing plastic, formed	700
Clear prismatic plastic	1150
Small-cell plastic louvers	925 lumens/sq ft
Brushed aluminum louvers	1150
White enameled louvers	1400
Gray enameled louvers	2300
Black enameled louvers	3500

For *uniform* patterns of luminaires that involve less than the full luminous ceilings indicated above, the allowable generated lumens/sq ft can be greater than that shown in the table.

Multiplier Factors (For the above generated lumens/sq ft):

100% Ceiling coverage	1.0
10% Ceiling coverage	4.0
1% Ceiling coverage	8.0
0.1% Ceiling coverage	16.0

NOTE: Because of their dominating influence in the visual field, luminous walls involve more critical brightness limits. Limit generated lumens/sq ft to 60–75% of the values indicated above for ceiling areas.

While a small area of brightness may be judged to be tolerable or even comfortable, a larger area of the same brightness intensity may be judged uncomfortable. As a practical consideration, it should be noted that large-area luminous elements (such as luminous ceilings, luminous walls, and window walls) require particular attention to the problem of adequate brightness control. Because these elements consume a relatively large portion of the normal visual field, they exert a more significant brightness influence and must therefore function within more restrictive brightness tolerances.

By the same token, a small glare source can be buffered by increasing the brightness of the background against which the source is viewed. Luminous elements which are sources of moderate discomfort and distraction in a low brightness environment may be quite innocuous in brighter surroundings because the contrast is reduced and the eye tends to adapt itself to the higher brightness background.

FIGURE 1–1.17

LIMITS OF VISUAL COMFORT (LOCATION)

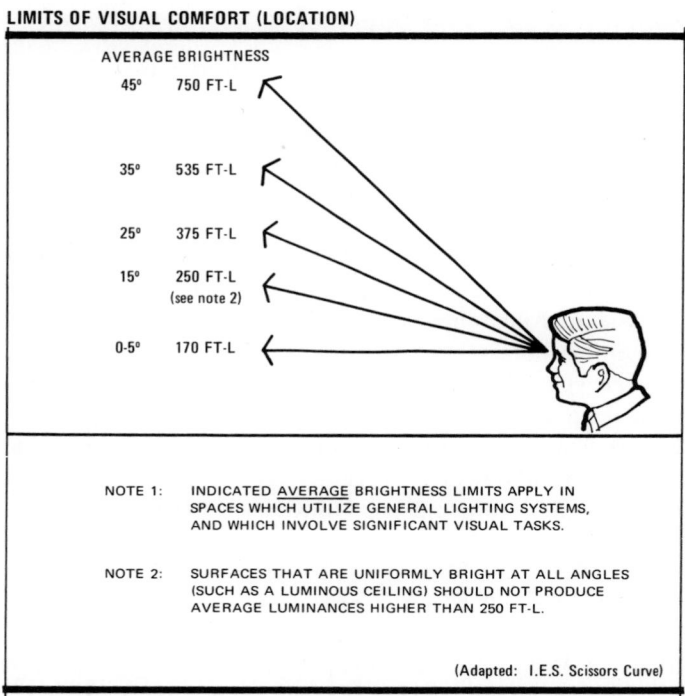

AVERAGE BRIGHTNESS

45°	750 FT-L
35°	535 FT-L
25°	375 FT-L
15°	250 FT-L (see note 2)
0-5°	170 FT-L

NOTE 1:   INDICATED AVERAGE BRIGHTNESS LIMITS APPLY IN SPACES WHICH UTILIZE GENERAL LIGHTING SYSTEMS, AND WHICH INVOLVE SIGNIFICANT VISUAL TASKS.

NOTE 2:   SURFACES THAT ARE UNIFORMLY BRIGHT AT ALL ANGLES (SUCH AS A LUMINOUS CEILING) SHOULD NOT PRODUCE AVERAGE LUMINANCES HIGHER THAN 250 FT-L.

(Adapted: I.E.S. Scissors Curve)

Brightness tolerance as a function of location

As a corollary to the relationship between intensity and area of brightness, the negative influence of a glare source depends upon its location in the normal field of view and its proximity to the central foveal area of the eye. Figure 1–1.17 illustrates typical average brightness levels that can be tolerated in different portions of the peripheral field of view. (*Maximum* brightness of relatively small highlight areas can be as high as three times the *average* brightness shown.) While these tolerances will vary somewhat with the state of adaptation of the occupant's eye, this diagram indicates that there must be increasing restriction of general brightness as the area in question approaches the center of the visual field.

These studies help to explain why brightness levels that are considered acceptable for luminous ceilings are found to be excessive and uncomfortable for luminous wall areas. Wall areas must function within more restrictive tolerances because they represent a more dominant influence in the normal visual field.

Adaptation and surround brightness

The subjective impression of visual comfort also depends on the brightness relationship between task surfaces and their surroundings. Facing a window with view of a bright overcast sky can make reading a book extremely difficult because of the effects of background glare. Equally difficult is reading a brightly illuminated book when the surroundings are in darkness.

In spaces where sustained visual work is involved (such as offices, classrooms, industrial areas, etc.), brightness relationships within the normal field of view should be controlled to allow the eye to adapt to an overall environmental brightness near the brightness of the task itself. In this way, the *shock effect* of high environmental contrast, as well as the strain of continual readaptation, can be mimimized.

In areas designed for prolonged work, therefore, research and experience have indicated the need for lighting the ceiling and walls (as well as the working surfaces) to avoid uncomfortable or fatiguing working conditions produced by excessive contrast. For comfortable seeing over a long period of time, the general brightness of surfaces immediately surrounding the task should not differ appreciably from that of the task itself. For work areas, it is generally recommended that spatial brightnesses average no less than 1/10 and no more than 10 times the average brightness of the task. (Also see subsequent discussion: "Effect of spatial context.")

Sparkle

If glare is an undesirable element in the environment, then the difference between *glare* and *sparkle* is an important design consideration. The principal difference lies in the relationship between brightness intensity and area, and in the prominence of that area in relation to the total visual field. If large areas of brightness are distracting and disconcerting to the viewer, relatively small areas of similar (or higher) intensity may be the points of sparkle and highlight that contribute visual interest and a sense of spatial vitality.

While the negative influence of glare is to be mimimized, then, this should generally be done without eliminating the closely associated stimulating influence of sparkle.

PERCEPTION OF DETAILED VISUAL INFORMATION

As implied in the previous discussion of the luminous background, the spatial lighting pattern should contribute to the identification of significant information centers—including task and work centers. But as soon as this identification has taken place for a given occupant, the suitability of the lighting on the task area itself will depend on its quality in assisting communication of precise visual detail. The remainder of this chapter will therefore focus on the affect of light on the specific ability to communicate a visual idea or message.

For purposes of this discussion, *communication* is defined as the act through which man derives the information necessary to gain knowledge to make an effective decision, or to perform a meaningful task. This process involves the use of the visual sense for observing situations and relationships, or for discriminating and differentiating precise detail. In this sense, the designer is concerned (1) with the general context within which the communication is perceived, (2) with the identification of general meaning, and (3) with the perception of meaningful differences.

Effect of Spatial Context

Brightness in the peripheral areas surrounding a specific, localized task center has an important effect on the ability to distinguish fine task detail (visual acuity). Optimum acuity is achieved when the general brightness difference between the central task (foveal vision) and the immediate spatial background (peripheral vision) is from 1:1 to 4:1, with the task area tending to be slightly brighter than the background. An increase in this ratio to 250:1 (task is brighter) will produce a reduction in acuity of approximately 10%; while for a bright task seen against a totally dark background, acuity is reduced approximately 20%. Some such loss in acuity would occur in a space such as that illustrated in Figure 1–1.18.

As a general rule, when highly precise visual performance is required, spatial brightness differences exceeding 10:1 should be kept well outside of the more central 40° visual cone. (But of course this restriction is not generally applicable for more casual visual activities where the drama of high contrast focal centers can make an important contribution in the *experience* of the space or activity.)

Even more significant differences in acuity occur when the spatial background is brighter than the task (i.e., the task detail approaches a silhouette condition). A relatively moderate 1:20 ratio (background is brighter) will produce a reduction in acuity of approximately 20%—and these reductions multiply rapidly as the background intensity increases. Consider the reduced visibility of detail in the silhouetted chairs in Figure 1–1.19.

Dark work surfaces seen against bright spatial backgrounds should

FIGURE 1–1.18
HIGH CONTRAST SPATIAL CONTEXT

FIGURE 1–1.19

FOREGROUND DETAIL IN SILHOUETTE

generally be avoided where precise perception of detail is required for effective visual performance or participation. (On the other hand, spatial silhouette may be useful when it is desired that the occupant be principally aware of general forms and spatial context, while deemphasizing his awareness of foreground detail. This is the so-called *cocktail lounge effect* that contributes to a sense of personal detachment or privacy. See previous discussion: "Spatial Distribution of Brightness.")

Task Contrast

For maximum perception of *detail* associated with the visual task, maximum contrast is desirable. In the immediate task area, the eye perceives detail through a sequence of quick eye movements that use foveal vision to scan the boundaries separating areas of different brightness and color. For example, white ink on white paper is imperceptible, while black ink on white paper provides contrast sufficient for the eye to separate the significant form and detail from the local background. For detail to be clearly definable against a background, then, there must be contrast between the two; and acuity improves as contrast increases.

In this sense, *brightness contrast* is defined as follows for diffuse surfaces:

$$\text{Contrast} = \frac{(\text{brightness of background}) - (\text{brightness of detail})}{(\text{brightness of background})}$$

If illumination is the same on both detail and local background (as it is for tasks such as reading from a page in a book), then the following definition applies:

$$\text{Contrast} = \frac{(\text{reflectance of background}) - (\text{reflectance of detail})}{(\text{reflectance of background})}$$

$$(\text{EXAMPLE}: \quad \frac{(80\% \text{ white paper}) - (20\% \text{ gray-black ink})}{(80\% \text{ white paper})} = 0.75 \text{ contrast})$$

In addition to *brightness contrast,* the eye also perceives detail and form through *color contrast.* Red ink is perceptible against its complementary color of green, even if both the red detail and the green background involve the same luminance (brightness). In practice, of course, most task contrast involves both brightness *and* color.

But these comments relate primarily to the contrast that is inherent in the task itself. There are also *environmental* influences that affect detail contrast. Generally, these relate to the effect of reflected brightness patterns in creating *veiling reflections* on or near a task—or the effect of shadows in reinforcing or complicating perception of three-dimensional form. There is also the influence of the lighting in modifying the spectral balance of the task.

These environmental influences will be discussed individually.

The effect of veiling reflections in reducing task contrast

Most surfaces reflect light both specularly and diffusely. Brightness produced by diffuse reflection depends on the intensity of illumination on the surface. Brightness of a specular reflection, on the other hand, is related to the brightness of the light source itself. This latter condition is a reflected mirror-like image of a lamp, of a luminaire, or of another reflecting surface—and this image, super-imposed on a glossy surface, will tend to obscure or *veil* the natural detail and color inherent in that surface.

In general, glossy surfaces should be avoided in the immediate vicinity of a significant local or spatial task. Glass-covered or highly polished desk tops, for example, have a very high specular component, and reflected images of overhead luminaires can become extremely distracting glare sources. Similarly, reflected images can disrupt the visual integrity of a polished wall surface. Where possible, then, low-gloss (mat) surfaces and finishes should be used for working surfaces and for major spatial elements (walls, floor, etc.).

As a corollary, specular reflections on a local work task (such as luminaire images superimposed on glossy paper) tend to reduce task contrast by obscuring printing and other relevant detail (see Figure 1–1.20).

FIGURE 1–1.20
VEILING REFLECTIONS

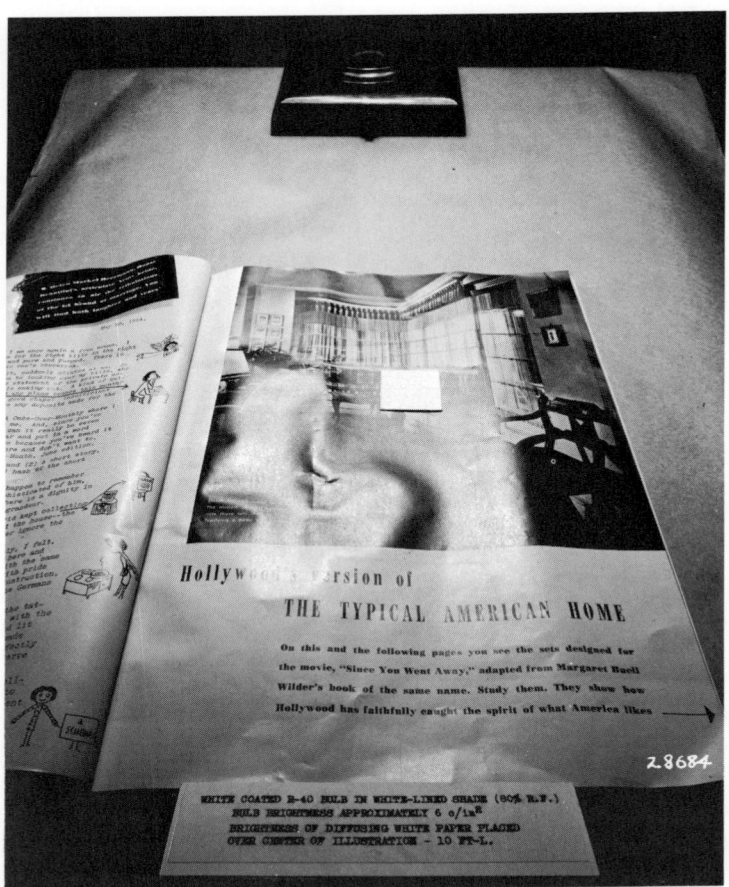

Where possible, such task finishes should again be modified to minimize the negative influence.

Elimination of veiling reflections on horizontal surfaces   When attempting to reduce the adverse influence of veiling reflections, the visual task or work surface may be tilted, or the luminaires may be located so that the lighting system is not a major source of veiling reflections. Both of these techniques involve a layout analysis that treats the work surface as a mirror—and all light sources, including bright walls and window areas, must be located outside of the *reflected field-of-view* (see Figure 1–1.21).

FIGURE 1–1.21

VEILING IMAGES (HORIZONTAL SURFACES)

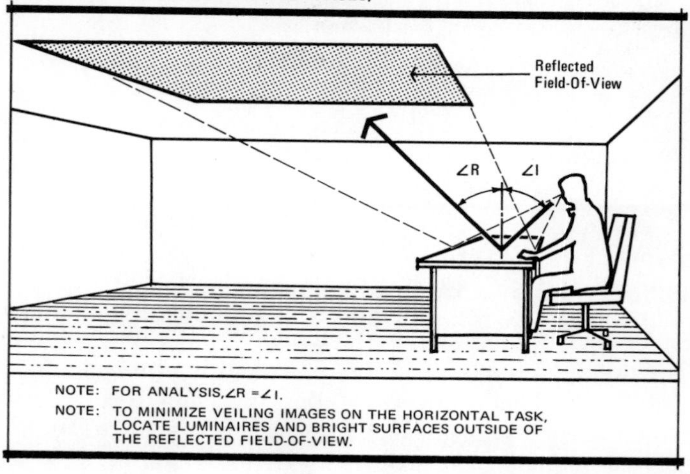

NOTE: FOR ANALYSIS, ∠R = ∠I.
NOTE: TO MINIMIZE VEILING IMAGES ON THE HORIZONTAL TASK, LOCATE LUMINAIRES AND BRIGHT SURFACES OUTSIDE OF THE REFLECTED FIELD-OF-VIEW.

Elimination of veiling reflections on vertical surfaces   Perception of vertical surfaces are also affected by veiling images, and this may further complicate the placement of lighting devices. As a general rule, low gloss or mat finishes are desirable where the visual integrity of the materials is to be preserved.

However, in some cases, a significant specular gloss may be inherent in the nature of the material—such as glass, marble, or high-gloss enamels and varnishes that are applied to facilitate ease of maintenance. When such glossy or polished surfaces cannot be avoided, the surface must again be analyzed as a *mirror,* and very bright elements should be shielded or removed from the *reflected field of view* (see Figure 1–1.22).

FIGURE 1–1.22

VEILING IMAGES (VERTICAL SURFACES)

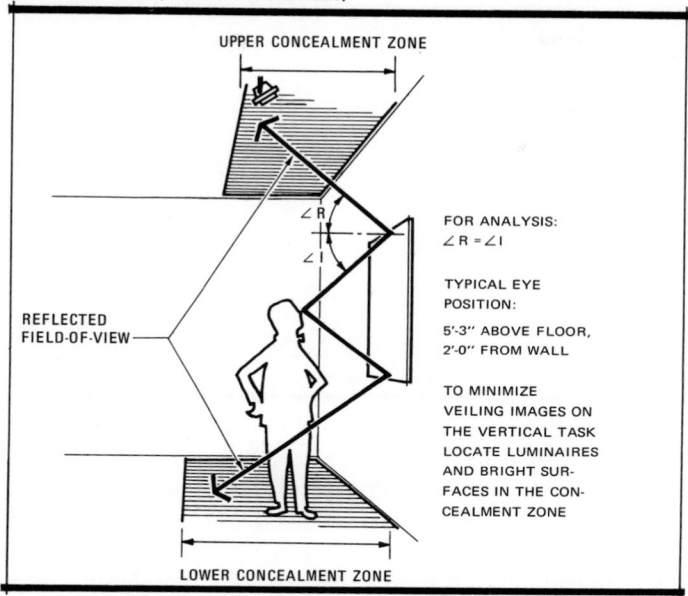

UPPER CONCEALMENT ZONE

∠R
∠I

FOR ANALYSIS:
∠R = ∠I

TYPICAL EYE
POSITION:

5'-3" ABOVE FLOOR,
2'-0" FROM WALL

TO MINIMIZE
VEILING IMAGES ON
THE VERTICAL TASK
LOCATE LUMINAIRES
AND BRIGHT SUR-
FACES IN THE CON-
CEALMENT ZONE

REFLECTED
FIELD-OF-VIEW

LOWER CONCEALMENT ZONE

Perception of color and color contrast

Colored materials reflect or transmit more strongly in certain regions of the spectrum, and the color that is perceived in an object or surface is determined by the characteristics of this transmitted or reflected light. This interdependence of light and color means that in order to provide accurate color rendition, the light source must emit those wavelengths which the object is able to reflect (or transmit). A deficient mixture will alter perception and cause the impression that specific colors are deficient or completely lacking.

For example, a green object under a red light source appears black or dark gray because the surface absorbs all colors except green, and little or no green is present in the red light to be reflected. In a similar way, any spectral deficiencies that are inherent in the prevailing light source will cause some surface colors to be *grayed*. This action tends to affect contrast adversely, and therefore reduces acuity.

As a result, the selected light source should generally produce energy in those regions of the spectrum that are meaningful in the task or in the decorative scheme.

FIGURE 1–1.23

PERCEPTION OF COLOR

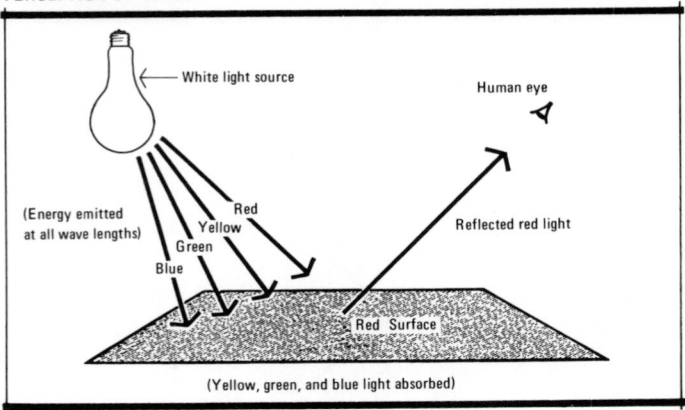

(Yellow, green, and blue light absorbed)

Detection of color differences   Differences in the spectral quality of *white* light can exert a decisive influence on acuity when the detection of subtle differences in surface color is involved (such as in color matching or color inspection). In most cases, *white* light that is rich in the spectral region where maximum *absorption* occurs (i.e., minimum reflectance) will tend to accentuate narrow differences in surface color (see Table 1–1.7).

Apart from precise inspection situations, these principles also underscore the important relationship between color of light and surface color when specifying fabrics, wall colors, and other elements in the spatial scheme. In the final specification, colored materials should be appraised, matched, and selected under the lighting condition(s) that will predominate in the actual space.

Table 1-1.7   Detection of Differences When Matching Colors

	Light Source Spectral Characteristics
Optimum detection of differences in blues, purples, blue-greens	Warm-toned source that is rich in red, yellow
Optimum detection of differences in reds, yellows, yellow-greens	Cool-toned source that is rich in blue, green

Appearance of human complexions   The spectral characteristics of the dominant light source are similarly significant in altering the appearance of human complexions. *White* sources that are rich in energy at the red end of the spectrum (filament, delux warm fluorescent) complement and flatter the complexion, imparting a ruddy or tanned character to the skin. *White* light sources that are strong in the yellow and blue ranges of the spectrum, but are weak in red (mercury and some fluorescent) tend to produce a sallow or pale appearance.

Shadow contrast and the perception of form and texture

Three-dimensional form is *seen* as a relationship of highlight and shadow. When this relationship is altered (by changing the directional characteristics of the lighting system), the impression of form may also be changed. In this sense, the lighting system should perform in sympathy with the inherent surface character of materials that define the physical space.

FIGURE 1–1.24
SURFACE FORM

Grazing light will emphasize highlight and shadow (see Figure 1–1.24). As a factor in visual acuity, this condition will improve perception of depth—which, in turn, may be important in the perception of texture or in the accurate performance of a very precise three-dimensional visual task. For example, grazing light will often aid in the perception of surface blemishes and errors in finish workmanship (see Figures 1–1.25A and

FIGURE 1–1.25A

SURFACE BLEMISHES: GRAZING
LIGHT

FIGURE 1–1.25B

SURFACE BLEMISHES: DIFFUSE
LIGHT

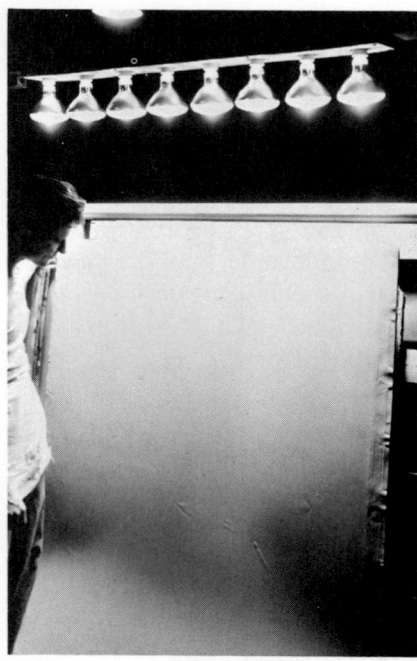 

1–1.25B). This, then, is a visual task that is representative of those that depend on direction of light rather than intensity or color to produce appropriate conditions for visual perception.

Conversely, a more diffuse lighting condition will reduce visibility of surface flaws and create a visual impression of flatness or surface uniformity that is more suitable for a plaster wall or tile ceiling. The diffusing effect may be the result of illumination emitted directly from the luminaire, or it may be produced by interreflection from other high reflectance surfaces (floor, walls, ceiling, etc.). In either case, the action of *frontal* light is particularly significant in reducing or eliminating the localized and random brightness variations that are due to subtle irregularities and flaws.

Texture and form   When the central visual task involves the perception of texture and three-dimensional form, this perception is influenced decisively by the character and distribution of light that impinges on the task.

The illustrations in Figure 1–1.26 suggest the considerable range of

FIGURE 1–1.26

INFLUENCE OF LIGHT DIRECTION ON TEXTURE AND FORM

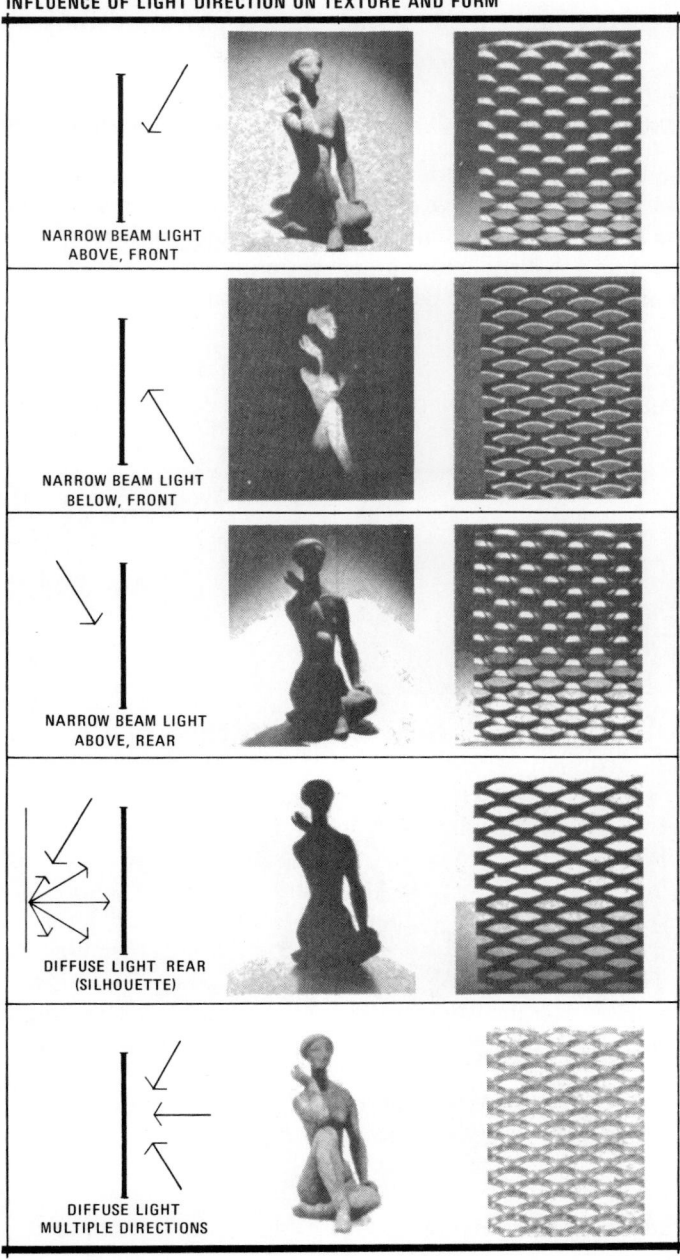

NARROW BEAM LIGHT
ABOVE, FRONT

NARROW BEAM LIGHT
BELOW, FRONT

NARROW BEAM LIGHT
ABOVE, REAR

DIFFUSE LIGHT REAR
(SILHOUETTE)

DIFFUSE LIGHT
MULTIPLE DIRECTIONS

visual impressions that result from changes in the directional composition of the light. They also suggest that analysis of *visual form* involves more than the physical form itself. Visual form is physical form modified by light.

Negative influences of shadow in the central communication area

For many common visual tasks, shadow may become an element of distraction in the immediate task center. In some cases, shadows on the work surface may be mildly irritating (such as those induced by the hand while attempting to write under a concentrating light source).

Where more precise and demanding visual study is involved, shadows

FIGURE 1–1.27A

ENVIRONMENTAL SHADOW IN THE WORK AREA

FIGURE 1–1.27B

ENVIRONMENTAL DIFFUSION IN THE WORK AREA

produced by this same concentrating lighting condition may become extremely disconcerting—and in some cases, hazardous in the sense that the shadows impede communication of visual information necessary for adequate safety (see Figures 1–1.27 A and B). The excessive concentration and constant readaptation required of a worker in these situations can, over a period of sustained work, result in visual fatigue, accidents, and errors.

However, this does not mean that highlight and shadow are never desirable in intensive working environments. Previous reference has been made to the advantages of grazing light in those instances that involve perception of three-dimensional tasks. Furthermore, just as highlight and shadow on a sunny day become emotionally stimulating visual influences, carefully placed brightness accents and shadow areas are useful for visual relief and interest in the interior environment. However, in most cases, a diffuse condition is desirable at the task center itself, and the effect of environmental shadows should be minimized in this area.

Detail Size

Most persons with normal vision can distinguish black detail on a white background if the detail subtends at least one minute (1′) of arc at the center of the eye. At a distance of 100 ft this indicates a typical minimum detail dimension of approximately 0.4 inches; at 1000 ft, a detailed dimension of about 4 in. is minimum. Where closer tasks are involved, a minimum of about 0.004 in. is the approximate minimum when the viewing distance is 12 in.

The relationship between brightness and size of detail

Optimum acuity is more easily achieved when the detail size exceeds this typical threshold condition. Increased detail size may facilitate relatively good visual communication even under somewhat adverse conditions. An example is the experience of relatively easy reading of a newspaper headline in a dimly-lighted room; while attempts at careful reading of smaller type under the same illumination becomes much more difficult and prolonged.

The nature of this relationship between size of detail and brightness is suggested in Figure 1–1.28. The indicated nomograph relationship exists for well-defined detail and form seen against varying reflectance backgrounds.

FIGURE 1–1.28

ILLUMINATION RELATIONSHIPS (TASK DETAIL)

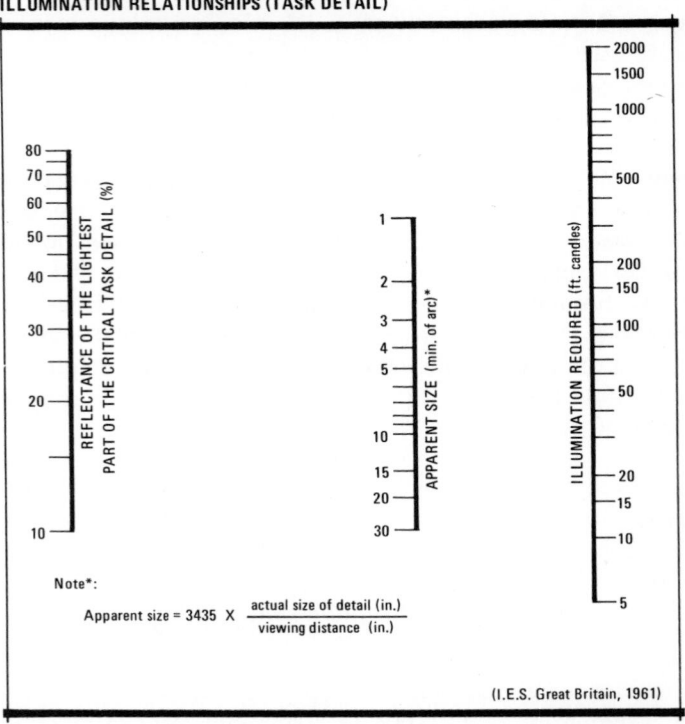

Note*:

Apparent size = 3435 X $\dfrac{\text{actual size of detail (in.)}}{\text{viewing distance (in.)}}$

(I.E.S. Great Britain, 1961)

Recognition and memory

Experience and memory are also factors in visual communication. Some forms are well-known to the observer and will be recognized under relatively poor visual conditions. This recognition is possible because the observer need not study the precise detail in order for a message or idea to be communicated to him. Commercial trademarks and traffic sign shapes are among the more obvious examples of this. In a similar way, quick recognition of many numbers, letter forms, and other common shapes depend on this principle.

Questions of size and contrast must therefore be evaluated in terms of the task. Is simple recognition satisfactory (as in reading)? Or is it necessary to study precise detail (as in many industrial tasks)?

Similarly, the character and quality of the light must be evaluated for its effect on visual recognition. Time for perception may be prolonged if the lighting is not *typical* or if the lighting does not provide the expected, remembered, or learned perception. For example, the act of searching for a red automobile can be prolonged under a mercury light source that is deficient in red energy.

Conditioned association   For the most part, this discussion of communication has been concerned with perception—and in this sense, it may be relevant to make some distinction between *perception* and *aesthetics*.

Communication of the *meaning* of what we see is based on *perception*. Is it round? Is it flat? Is it an orange? Is it an apple? Judgment or *value* is based on *aesthetics*. Is it a good orange? Is it a pleasant texture? Is it an appropriate space?

Often these value judgments are based on previous experience—on memory of the meaning of a similar perception or idea. Some of these associations are related to the nature of the light that impinges on the detail, and in this sense the lighting system may exert a direct influence on aesthetic judgment.

Artists have long been aware of these variations in our perception of light and have skillfully utilized the psychological associations which these variations induce. In painting, for example, the contrast between *yellowish* highlights (from direct sunlight) and *bluish* shadow (illuminated by skylight) is sometimes exaggerated to achieve a *natural* impression.

In a similar way, a lighting condition that alters or reverses the usual direction of light changes the normal (natural) relationship of highlight

FIGURE 1–1.29A
LIGHT FROM A NATURAL DIRECTION

FIGURE 1–1.29B
FROM AN UNNATURAL DIRECTION

and shadow. Aesthetic judgment may therefore be based on an unnatural perception, possibly inducing a sense of uncertainty, mystery, or even fear (see Figures 1–1.29A and 1–1.29B). While most situations probably favor the development of more natural environmental situations, special conditions can be visualized where the environmental condition is manipulated to produce a complementary aesthetic impression (such as the sense of mystery produced by lighting a Buddha or a Gothic gargoyle).

In cases that involve perception and judgment of objects and forms, then, variations in light (color of light as well as direction) will affect the observer's unconscious judgment of what he is seeing. Again, judgment is based on perception of the physical form *as modified by light*. But remembered associations may be decisive in the communication of an idea or experience.

Brightness of Typical Tasks

Research indicates that increasing light energy (flux) is generally required to maintain a constant level of acuity when the following conditions occur:

(1) as the size of the detail is reduced
(2) as the contrast between the detail and the background is reduced
(3) as the task reflectance is reduced
(4) as the time permitted or utilized for perception is reduced

As a generality, then, (and assuming constant contrast, size, and time for viewing) visual acuity increases with brightness. There is a particularly high rate of improvement when low initial intensities are involved. This generally reflects the increasing influence of *cone vision* over *rod-dominated vision* as brightness intensities increase from minimum conditions. Once the cones begin to approach full stimulation, then, acuity continues to improve as brightness increases, but at a significantly slower rate of change (see Figure 1–1.30).

The influence of light intensity

For typical central visual activities or tasks in which neutral-colored two-dimensional form or detail is viewed against a high reflectance background, a high level of visual performance is obtained with approximately 50 ft-c of incident illumination. With this type of task, performance losses or gains attributable to moderate changes in light intensity are relatively small (see Figure 1–1.30).

Table 1-1.8 Illumination Relationships (Typical Intensities Required to Facilitate Equal Visibility)

footcandles	Office-Type Tasks		Industrial-Type Tasks	
1	Typed original, good ribbon	1.0 ft-c		
	Text material, well printed	1.1 ft-c		
	Price tag, ink	3.1 ft-c		
5	#2 pencil line on tracing, over blueprint	6.6 ft-c		
10			New micrometer, etched	7.4 ft-c
			Tailor's mark: white chalk, blue cloth	10 ft-c
			Tailor's mark: white chalk, gray cloth	21 ft-c
50	#2 pencil writing, poor quality	63 ft-c		
	#3 pencil, shorthand copy	76 ft-c	Darned blemish, gray cloth	74 ft-c
100	Typed copy, 5th carbon	133 ft-c		
	Price tag, pencil	241 ft-c		
500			Lumber inspection: seasoning checks, from 30 in.	317 ft-c
1000	Thermal reproduced copy, poor quality	589 ft-c	Vernier calipers, nonetched	631 ft-c
			Pocket stitching, tan shirt	1890 ft-c
5000	Typed original, extremely poor quality	3140 ft-c		
	Tracing over blueprint	5090 ft-c		
10,000			Tailor's mark: white chalk, tan cloth	10,000+ ft-c

(Adapted: I.E.S. United States, 1966)

Table 1-1.9 Task Surface Illumination Levels (Typical Recommendations)

	Casual Visual Activity 0–30 ft-c	Moderate Visual Activity 30–75 ft-c	Extended Visual Activity 75–150 ft-c	Difficult Visual Activity 150–250 ft-c
GENERAL				
Circulation:				
Corridors, escalators, elevators, stairways	●			
Lobbies		●		
Storage:				
Locker-, toilet-, and wash-rooms	●			
Inactive; active (rough, bulky items)	●			
Active (medium-sized items)		●		
Active (small-sized items)			●	
OFFICE				
Office work:				
Reading, transcribing, filing			●	
Accounting, auditing, tabulating, business machine operation			●	
Cartography, design, drafting				●
Conference:		●		
SCHOOL				
Classrooms:				
Classroom work, library, study		●		
Manual arts, drafting			●	
Sewing rooms, laboratory benches				●
Assembly:				
Auditoriums, cafeteria, gymnasiums (exercise)		●		
Gymnasiums (exhibition games)			●	
STORE				
Service areas:				
Merchandising areas			●	
Showcases, displays				●

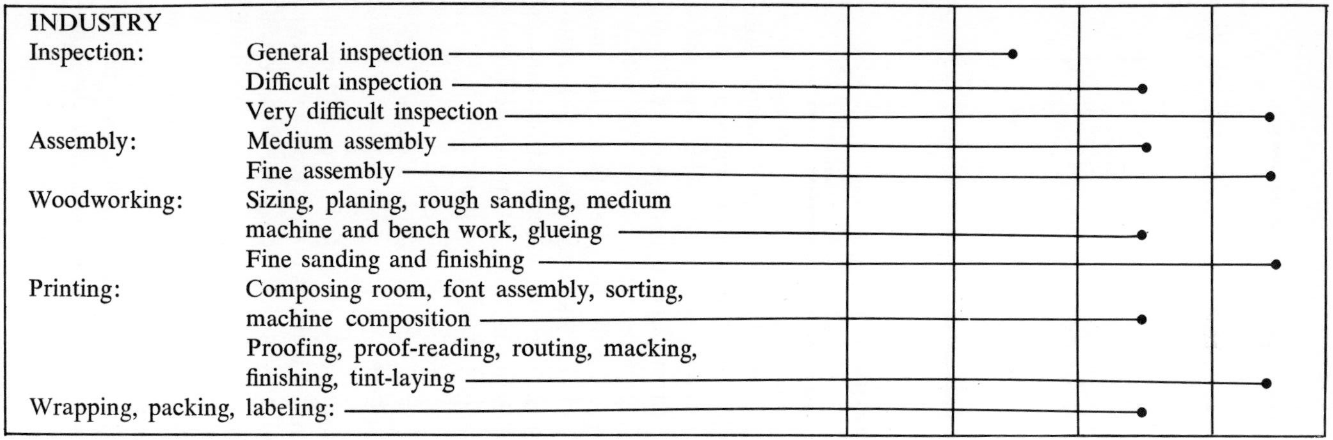

INDUSTRY

Inspection:
General inspection
Difficult inspection
Very difficult inspection

Assembly:
Medium assembly
Fine assembly

Woodworking:
Sizing, planing, rough sanding, medium
machine and bench work, glueing
Fine sanding and finishing

Printing:
Composing room, font assembly, sorting,
machine composition
Proofing, proof-reading, routing, macking,
finishing, tint-laying

Wrapping, packing, labeling:

FIGURE 1–1.30
ILLUMINATION RELATIONSHIPS (VISUAL ACUITY)

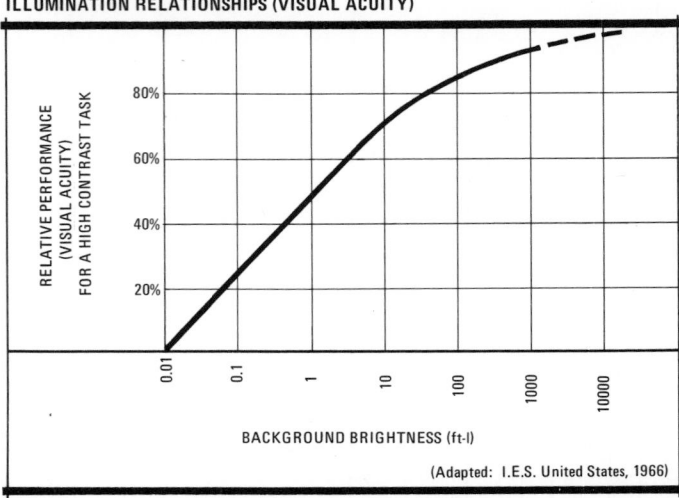

(Adapted: I.E.S. United States, 1966)

However, when the same detail becomes grayed or is viewed against a gray background (both effects reduce contrast), visual performance follows a much lower curve. In this case, variations in the intensity scale are quite significant over a wider range; and very much higher intensities may be required in order to achieve near-maximum perception of detail.

Illumination intensity requirements therefore vary with the difficulty of the work task, as determined by the previously discussed factors of size, time for exposure, and contrast. Table 1–1.8 indicates the relative intensities of light required to be incident on various representative tasks in order to produce an approximate equal level of acuity when diffuse illumination is provided (i.e., no specular *veiling* images are present).

Recommended light intensity ranges

Recommendations for illumination intensity on a task are, at present, generally published in terms of *footcandles* incident on a typical task area. For most commercial, industrial, and educational activities, this typical work surface is taken to be a horizontal plane 30 in. above the floor. (This is desk or table height.)

General recommendations for representative areas and activities are indicated in Table 1–1.9. These recommendations and the other detail-oriented factors discussed in the latter part of this chapter should be developed concurrently with the relevant background considerations discussed earlier.

The Sonic Environment

HEARING

The sense of hearing is based on the ability of the ear to selectively process a range of airborne vibrations. This sense is primarily used for communication of information through such devices as verbal dialogue or warning signals. Again (as in vision), there is a processing of simultaneous or successive information cues and an assigning of mental priorities regarding importance.

Hearing is often the decisive sense for precise communication of ideas between individuals, and is particularly dominant when visual conditions are poor.

FIGURE 1–2.1

THE HUMAN EAR

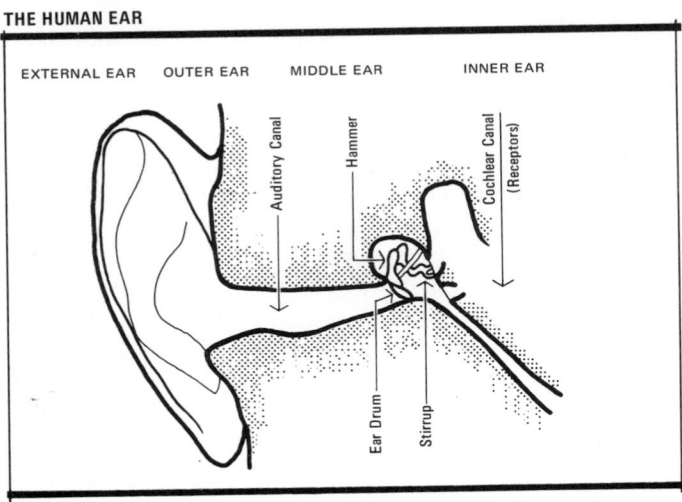

EXTERNAL EAR   OUTER EAR   MIDDLE EAR   INNER EAR

Auditory Canal

Hammer

Cochlear Canal
(Receptors)

Ear Drum

Stirrup

Spatial orientation may also be affected, in that there will likely be an instinctive desire to identify and locate the source of sonic signals. Furthermore, there may be a subjective sensation of spatial *hardness* or *softness* communicated through the sense of hearing.

Sonic Sensation

The human ear detects sound over an extremely broad range of intensities and frequencies by responding to small variations in air pressure. These variations induce a subtle vibration of the ear drum; and this, in turn, induces variations within the liquid-filled cochlear canal, where the true auditory receptors are located. In terms of time separation, two clicks as near in sequence as 0.001 sec can be detected by the ear as separate signals.

However, the ear does not perceive the whole range of vibration frequencies as sound. For example, a tuning fork vibrating at a rate of 15 cps arouses no sensation of hearing. Most people do not perceive sound until approximately 30 cps is reached.

The full range of human response to sound involves a frequency *spectrum* that ranges from approximately 30 cps at the low end, to approximately 10,000 cps at the upper end. For some individuals, the range of sensitivity extends as low as 20 or as high as 20,000 cps.

Optimum sensitivity occurs in the frequency range between 500 and 6000 cps. This range includes most of the principal *speech range.*

The sonic spectrum and the area of audible sensation

The intensity of sound is measured in *decibels,* and the listener's acuity (i.e., his ability to hear detail) is measured in terms of the threshold decibel levels at which he can detect sound signals at various frequencies.

The variable subjective sensation of sound intensity is summarized for pure tones in the form of *equal loudness contours* (see Figure 1–2.2). These curves generally define the area of audible sensation.

The 1000 cps tone is the base intensity for each of these curves, and the curves indicate the intensity levels at each frequency that will be subjectively interpreted by a normal listener to be the same loudness as the 1000 cps tone.

The threshold of hearing

0 db on the sound intensity scale is, by definition, the *natural* threshold of hearing at 1000 cps. Most normal listeners can detect a slightly lower

FIGURE 1-2.2

EAR SENSITIVITY (EQUAL LOUDNESS CONTOURS)

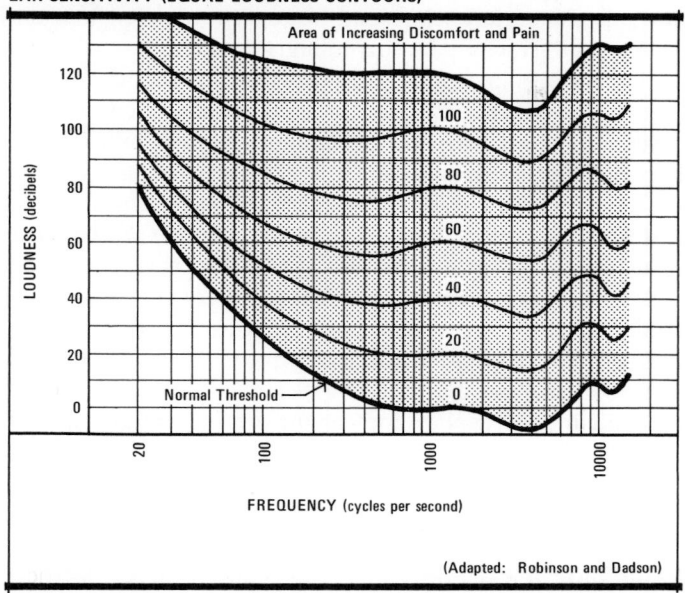

(Adapted: Robinson and Dadson)

intensity signal at 4000 cps; while considerably higher intensity levels are required to produce minimum perception at both ends of the frequency spectrum.

The threshold that is defined in this manner assumes an idealized *quiet* background. However, when background noise is present, a somewhat higher *artificial* threshold is created. For example, while a 10 db signal at 1000 cps would be easily perceptible for normal human listeners in a totally quiet environment, a background noise level of 25 db would establish an artificial threshold that would eliminate perception of the original 10 db sound. In order to be perceived by the same listeners, meaningful signals must now exceed the 25 db background.

Distortion of high intensity signals   When background noise is very intense, it becomes quite difficult to communicate an intelligible signal. This is due, in part, to the high level artificial threshold. But it is also due to a condition within the ear that is somewhat analogous to *glare* in the eye.

In this regard, efforts to increase the intensity of communicating signals (by shouting or by using electronic amplifiers) must be limited because a very high intensity signal will tend to produce interior distortions within the ear itself. These distortions will, in turn, obscure the clarity of the signal by introducing an internal masking effect.

When low or moderate background noises exist, then, improved communication can be facilitated by increasing the signal intensity. But when high levels of background noise are present in the environment, the effectiveness of this technique is limited, and an improved communication condition will more likely depend on methods which reduce the artificial threshold created by the noise.

Pain and hearing loss   The action of the hammer in the ear provides some muscular adjustment of the ear drum to prevent injury at high intensity levels. But exposure to sound pressure levels in excess of 120–130 db will generally cause increasing discomfort and pain in the ear. Such intensities may also produce momentary disability, or even more permanent sensory damage.

This region defines the upper limit of the area of audible sensation, and is called the *threshold of discomfort and pain.*

An abrupt but less intense noise (such as the noise produced by a large exploding firecracker) will produce momentary deafness. Although individuals differ considerably in their vulnerability to such effects, the intensity and length of deafness will generally depend on the intensity and duration of the infringing, high intensity sound.

Permanent hearing loss can develop from repeated or continuous exposure to moderately high noise levels. For example, extensive exposure to levels in excess of 85 db may cause permanent losses in sonic sensitivity (see Table 1–2.1).

And of course, hearing loss can also occur due to normal aging processes. As an indication of this deterioration in hearing with age, one study found an average sensitivity loss among men in their sixties to be about 30 db. Among men in their seventies, the loss rose to 40 db or more.

Regardless of cause, in most cases, the revised natural threshold that results from permanent losses in the ability to hear is most marked in the higher frequency regions of the spectrum. This will, of course, eliminate the affected listener's perception of lower intensity signals in this region.

Multi-channel listening

The two ears can be used together when a single dominant signal is involved. Or, to some extent, they can be used as separate channels when several sources or signals are heard at the same time.

If the information rates are not too high, separation of sound sources can be easily accomplished by the listener without need for external modification of the signal itself. However, as the complexity and multiplicity of signals become greater, the listener will find that he must rely on one (or more) of three methods of separation:

(1) He can develop separation in amplitude (generally by moving closer to one sound source).

(2) He can develop separation in frequency (such as a conscious or sub-conscious decision to listen only to the lower frequency voice and disregard the higher frequency sources).

(3) He can develop a separation in space (possibly by turning the head so that one source is dominant on one ear; while all other sources are acting on the other ear, which can be rejected).

Spatial judgments   While sonic spatial judgments are crude, multi-channel listening provides a limited auditory *perspective* that facilitates judgment of distance and direction.

Judgment of *distance* is relatively simple, although not very accurate or precise. This may involve loudness or volume; it may also involve the complexity of the sound. In this latter case, the full-frequency modulation of a nearby sound becomes a simpler and more limited sonic spectrum as distance increases.

Directional judgments are of two types: those that require one ear, and those that require two.

For example, some judgment of sonic direction may be subconsciously based on previous knowledge, such as the association of traffic noise with the direction of a nearby road. Similarly, visual cues take precedence over sonic cues, so loudspeaker sounds are easily associated in direction with a movie screen or a lecturer. These examples represent readily localized sound sources and require only one ear for adequate hearing (or two ears working together).

When previous knowledge and vision are both eliminated, however, a sense of direction is developed by using each ear separately. Generally, this is accomplished by movement of the head so that each ear is stimulated somewhat differently (see Figure 1–2.3).

The Sonic Environment

In summary, the spatial patterns that are involved in human hearing are generally based on sonic contrast. If a meaningful sound falls within the normal response range, but is obscure due to lack of intensity or lack of contrast with the background, listening becomes difficult or seemingly irrelevant. On the other hand, the sensitivity of the ear is limited in the sense that meaningful signals can also become too loud for pleasant and

FIGURE 1–2.3

DIRECTIONAL JUDGMENTS (SONIC)

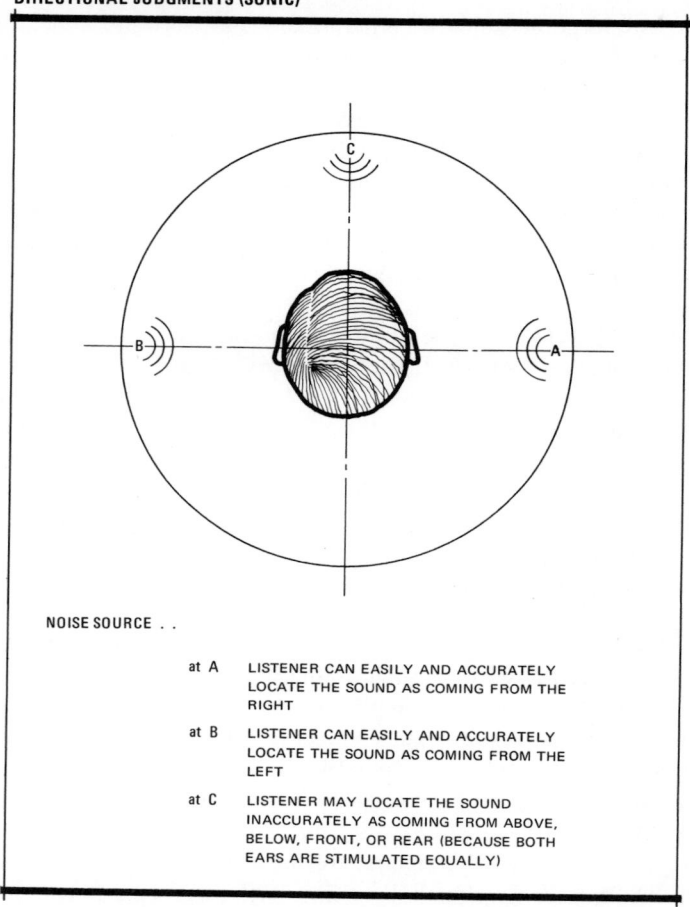

NOISE SOURCE . .

at A   LISTENER CAN EASILY AND ACCURATELY
        LOCATE THE SOUND AS COMING FROM THE
        RIGHT

at B   LISTENER CAN EASILY AND ACCURATELY
        LOCATE THE SOUND AS COMING FROM THE
        LEFT

at C   LISTENER MAY LOCATE THE SOUND
        INACCURATELY AS COMING FROM ABOVE,
        BELOW, FRONT, OR REAR (BECAUSE BOTH
        EARS ARE STIMULATED EQUALLY)

comfortable listening. Both of these extremes introduce adverse influences that can impede effective participation in an activity.

Excessive or irrelevant background noise may also obstruct an activity by introducing an element of distraction. The intensity of this distracting influence will depend on the tolerance and sensitivity of the individual listener; and this may vary for the same individual at different times.

In a still more nebulous area, sonic judgments may relate directly to the individual's ability to orient himself and to comprehend the nature of his environment. He uses his sense of hearing to make crude judgments of distance and direction when vision is lacking or obscured. He also perceives a *sense of hardness* or *softness* that may be associated with a room.

Design of the sonic environment is therefore primarily concerned with two aspects of control: (1) the *communicating signal,* and (2) the *sonic background.* The balanced manipulation of these sonic conditions should provide for the listener's need to receive communicated ideas by distinguishing relevant signals as separate from the general background. At the same time, this environmental balance should reflect the need to protect the individual occupant from excessive intensity levels, and from irrelevant signals that may confuse his sense of orientation or distract attention from his primary activity.

PERCEPTION OF THE SONIC BACKGROUND

Subjective sensation of the sonic background is primarily a function of (1) reverberation, and (2) random noise intensity. Both of these have some influence on the occupant's sense of spatial appropriateness and well-being. Both will also influence his ability to transmit and receive a sonic communication.

Reverberation

When a sound source ceases to emit energy, the direct sound stops immediately. However, the energy within an enclosed space will continue to reflect between the room surfaces; and as these waves successively pass the listener's ear, the original sound will continue to be heard at diminishing intensity for a very short period of time after the source itself has stopped. This subtle prolongation of the sound in the room due to continued multiple reflections is called *reverberation.*

There are spatial connotations associated with reverberation. A common experience is the perceptual response to a hard, empty room compared with the spatial sense of the same room after it has been softened by upholstered furnishings or by groups of people.

When soft, porous materials predominate, sound reflection is significantly reduced and internal sounds fade very rapidly. This *short reverberation time* (i.e., a rapid rate of fade) produces a spatial condition that we describe as acoustically dead. When room surfaces are consistently hard and highly reflective, on the other hand, sound continues to reflect in an unimpeded fashion between these surfaces and therefore fades more slowly. The effect of this type of space (a *high reverberation time*) would, in turn, be described as a live space.

FIGURE 1–2.4

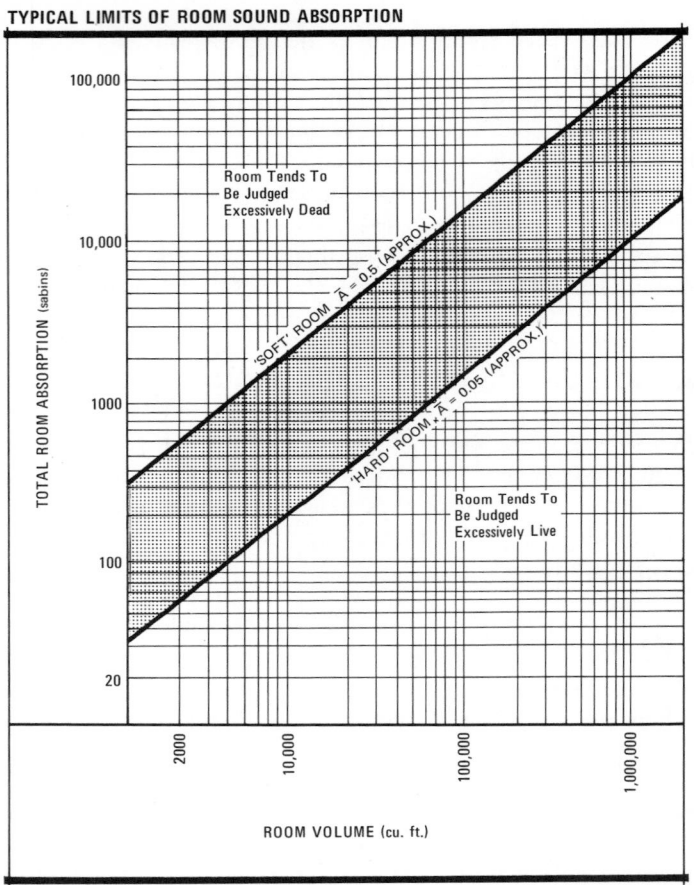

TYPICAL LIMITS OF ROOM SOUND ABSORPTION

Spatial absorption

Figure 1–2.4 is useful for estimating the spatial deadening effect of room absorption.

When the total absorption value of room surfaces and occupants indicates a position in the lower portion of the shaded area, the space tends to be perceived as *hard* or *live* by the occupants. If the total absorption lies below the lower limit line, most occupants can be expected to judge the space as *excessively hard* for normal activities. Above the upper limit line, the room may be judged excessively *dead*.

(Also see Figure 1–2.8 and subsequent discussions of room absorption in Chapter 2–2.)

Perceptual response to background noise

Background noise is most often discussed within the context of sonic communication. For example, it is sometimes roughly stated (as a rule of thumb) that background noise levels of 0–60 decibels will permit relatively easy conversation. With background noise levels above 60 db, conversation becomes increasingly difficult; until, at about 115 db, sonic communication tends to become impossible. At this upper level, distortions within the interior of the ear will tend to mask further attempts to communicate intelligibly with high intensity signals (such as by shouting into the ear of the listener).

But beyond those implications associated with communication, moderate and high intensity background noise may also influence the perceptual response to a space utilized for *noncommunicative* activities.

The effect of background noise on noncommunicative performance

While *pure tone* (single frequency) background noises are generally somewhat disruptive in the environment, *white noise* is a more manageable influence.

A *white noise* refers to a broad spectrum distribution of sound energy. With energy emitted at all frequencies, a typical white noise might be that associated with air escaping from a jet. When this energy is distributed somewhat uniformly over a very wide band, it produces a *hishing* sound. When the broad band noise shifts toward higher frequencies, it becomes more of a *hissing* sound.

There have been some studies of the effect of such background noises on individual performance when auditory communication is not involved in a major way. In this respect, background noise should be divided into two categories: (1) the continuous white noise and (2) the intermittent white noise.

The continuous white noise   Unless the intensity is sufficient to cause pain or permanent damage, continuous white noises appear to produce no measurable disturbing effects, and they apparently have little effect on participation in noncommunicative activities. Individuals can apparently adapt easily to moderate intensities of middle-frequency background noise.

There are some exceptions to this general rule, however, when higher background intensities are present and the occupant is confronted with high levels of information input (for example, when he is involved in a highly complex task, or one in which considerable vigilance is required). In this situation, background noise may produce a measurable reduction in performance and reaction time. However, when the occupant's tasks are simplified, an adverse effect on performance is no longer measurable in the same sonic environment.

Since moderate levels of continuous white noise can be present without creating a disruptive influence, then, a selected background noise can be utilized or applied to establish a moderate artificial threshold. This is often useful for masking objectionable lower intensity sounds—particularly in spaces where the need to communicate aurally is not a decisive consideration. (Also see subsequent discussion of "Supplementary Noise Screens and Masks.")

Physical Tolerance to High Background Noise Intensities However, continuous background noises must be handled carefully, because higher noise levels can produce permanent impairment of hearing.

In this regard, individuals vary considerably in their susceptibility to hearing impairment. However, it has been observed that at an early stage of the problem, damage is particularly marked at approximately 4000 cps. As a result, this area of the spectrum offers a potential test for early identification of those who are particularly sensitive to hearing damage.

When the problem is discovered, the individual involved may be transferred to a more quiet environment; he may make use of external devices, such as ear-plugs or ear muffs; or environmental changes can be made to reduce the noise to a more tolerable level.

In attempting to define the general conditions where impairment problems may be expected to occur, some preliminary studies have been based on the criteria that after 10 years of daily exposure, an individual should not have suffered any appreciable impairment of ability to understand speech at normal voice levels. These studies have produced a tentative conclusion that for an 8-hour working day, a sustained level of 85 db for each of the octave-bands above 700 cps is the approximate limit that can be tolerated over a long term without producing permanent damage to hearing (see Table 1–2.1).

The intermittent white noise Abrupt or significant changes in background noise conditions (both increasing and decreasing intensities) can produce a startle response or reflex action that involves muscle contraction, blinking, etc. Essentially, this is explained by the fact that the listener's attention is shifted suddenly to a new information source.

As the spectral band varies from the white noise and becomes narrower and more concentrated in frequency, the intermittent background noise begins to take on some aspects of pitch. As this occurs, the occupant's reflex response will tend to become more intense and pronounced.

Table 1-2.1 Background Noise Intensities (Tolerance)

Daily Exposure (Sustained Period)	Intensity Limit to Prevent Permanent Impairment of Hearing in the Speech Range
8 hr	85 db max.
4 hr	88 db max.
1 hr	94 db max.
½ hr	97 db max.

NOTE:

80 db range:	Automatic appliances, such as washing machines, dryer, etc.; inside typical residence with heavy auto traffic within 50 ft.
90 db range:	Typical industrial area; typical cooling tower from a distance of 20–30 ft.
100 db range:	On street with heavy truck traffic; noisy industrial area; typical mechanical equipment room.

While these environmental variations have the potential negative effect of distracting occupant attention from his current focus, the listener normally has the ability to quickly derive considerable information as he searches for and identifies the new sound source. This characteristic has obvious intrinsic value as a warning device.

Related to this, moderate variations in a white noise background can be effective in relieving boredom. While the momentary ability to derive very high quantities of information tends to diminish after the initial reflex action, the occupant tends to be more alert and responsive in the sense that his rate of information assimilation will generally level off at a higher plateau than that which immediately preceded the response.

Annoyance

In the intensity range that lies above the lower physical limit of threshold audibility but below the limit of threshold pain, it appears that an individual's psychological tolerance for background noise in noncommunicative situations will depend: (1) on his conditioning; (2) on his ability to maintain his concentration at a given point in time; and (3) on the information conveyed by the sound.

In the latter category, there is evidence that noises which mystify the occupant are more likely to become annoying than are sounds that can be easily located and identified by the listener. Intermittent or irregular sounds are more likely to be distracting than steady or continuous sounds. High frequency pure tones are more likely to be distracting than lower frequency tones or broad-band white noises. Furthermore, noises that seem to be avoidable, unnecessary, or inappropriate for the activity are generally found to be particularly distracting.

PERCEPTION OF SONIC SIGNALS

Sound, like light, is a medium for communication of information and ideas.

Perception of meaningful sonic signals will of course depend on the emission characteristics of the source and the proximity of the source to the listener. But for a given source *spectrum,* the listener's threshold of sensitivity may obscure some portions of the emitted sound. This threshold condition may be a *natural* one, i.e., the ear may be insensitive to some lower intensity sounds. It may also be an *artificial* threshold, i.e., the background noise level may obscure or screen lower intensity signals.

When either of these subthreshold conditions occurs, the listener will fail to perceive the affected signals. To that extent, the listener will suffer a loss of sonic comprehension, and a breakdown in communication may occur. Ideally, then, the sound contrast associated with communication activities must be sufficient to facilitate full perception and differentiation of sonic detail.

Voice Communication

Voice communication (speech) involves a great variety of frequencies and intensities. Each letter or syllable is enunciated as a characteristic sound, and comprehension depends on the listener's ability to perceive each signal and distinguish it from other similar, but subtly different signals.

When only 90% of the words are heard correctly, voice communication (listening) can become very fatiguing; and below this point, sonic communication becomes nearly impossible if accuracy or speed is required. An individual who is familiar with the language should be able to comprehend about 97% of the words for easy listening.

Speech signals

Although the majority of voice signals fall within the frequency range of 500–3000 cps, optimum perception of English speech involves a total *speech range* of approximately 200–6000 cps.

Voice signals emitted in normal speech also vary significantly in intensity—exhibiting a variation of approximately 25–35 db from the faintest to the more intense sounds.

Vowels are generally lower frequency signals of relatively high intensity and long duration. Consonants are higher frequency signals of lower intensity and shorter duration. The consonants generally have the greatest effect on communication because they contribute most of the actual *information* required for comprehension. Unfortunately, because they are lower intensity signals, consonants are also most susceptible to masking by background noise.

Intensity variations due to voice level and distance Figure 1–2.5 defines the approximate range of sounds that are emitted in speech. However, the intensity limits indicated in the diagram will vary with the voice level of the speaker. As shown, the *conversational speech range* approxi-

FIGURE 1–2.5

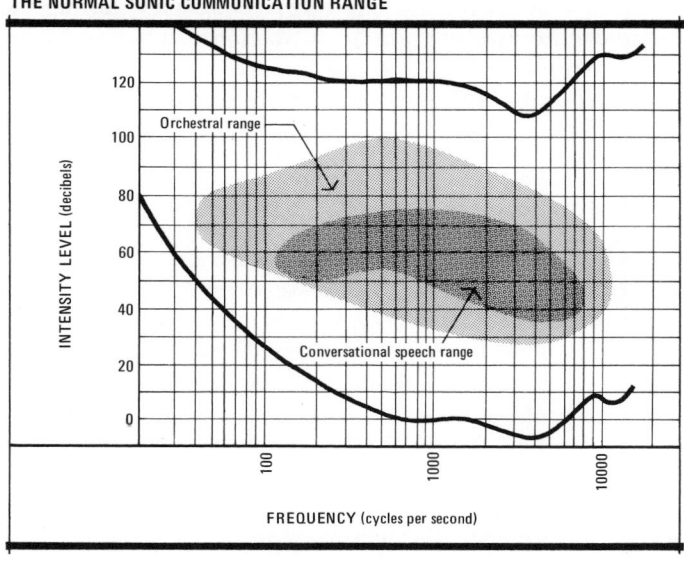

THE NORMAL SONIC COMMUNICATION RANGE

mates the signals perceived by a listener standing about 3 ft from an individual who is speaking in a normal voice.

When necessary, an increase in intensity can be produced by changing the intensity of the signal itself, or by changing the source-to-listener distance.

If the speaker lowers his voice to a more intimate level, the entire *speech range* moves down about 6–7 db (i.e., all speech sounds become fainter in approximately equal proportions). If the speaker raises his voice, the speech range rises about 6–12 db above the standard indicated in the diagram. If he shouts, the range moves up another 6–7 db.

Direct signal intensity also varies with the distance from the speaker to the listener. For each doubling of the distance (from the 3 ft standard previously noted), the speech range is reduced approximately 6 db. The converse is also true; for each halving of the distance, the range rises approximately 6 db.

The sense of sonic territoriality   An individual's sense of territoriality subjectively expresses the spatial volume that surrounds him. It helps to define his sense of relationship with other occupants.

The precise limitations of this sense of *personal space* will vary with cultural background and with the specific activity involved; but this influence is generally present in some degree and tends to define the proximity to another person that an individual will find acceptable and appropriate under a given set of circumstances. Inappropriate territorial relationships or environmental conditions that produce unnecessary infringements on personal space can become a cause of distraction or concern among more sensitive occupants.

Table 1–2.2 represents the general nature of this psychological relationship as it helps to define appropriate signal intensities for North Americans.

Multiple Signal Vectors

When sound travels from a source to the listener by a single path (as in normal conversation at close range), relevant variations in the signal are generally clear and distinct. However, when greater distances are involved, the sound may take several paths to the listener: (1) the direct path; (2) reflection from one or more room surfaces; and (3) the possible action of one or more loudspeakers.

Sound moves relatively slowly through air (approximately 1140 ft/sec). If the various major paths are significantly different in length, then, hearing problems may occur because the various vectors of the signal arrive at the listening location at significantly different times.

Table 1-2.2 Sonic Territoriality

Appropriate Audible Signal	Typical Intensity Range (Source)	Sense of Personal Space	Physical Proximity	Implies
Soft whisper	—	Very close	3 to 6 in	Top secret communication
Audible whisper or intimate voice	44–69 db	Close	8 to 20 in	Confidential communication
Normal voice	50–75 db	Neutral	20 to 60 in	Personal communication
Loud voice	56–81 db	Public (near)	5½ to 8 ft	Nonpersonal communication or group information
Overloud voice	62–87 db	Public (across room)	8 to 20 ft	Group address
Shouting	68–93 db	Upper limit	Over 20 ft	Hailing

If the various signal vectors arrive nearly simultaneously (i.e., an arrival time difference of 0.035 sec or less for speech), the stimuli will *reinforce* each other because the arrivals are closer than the time required for the source (speaker) to emit the next distinct signal. However, if the individual vector arrivals exceed the 0.035 sec spacing, the resulting interference can begin to reduce intelligibility.

When the space is small and normally proportioned, interfering reflections of this latter type are generally not significant. It is in larger rooms, where there are multiple reflections from more remote room surfaces, that problems of signal muddling and echoes may occur. (Distinct echoes will occur when the listener hears a sufficiently intense reflected sound 0.06 sec or more after he hears the direct sound.)

The sense of spatial intimacy

For communication activities that take place in moderate to large rooms (such as those intended for presentation of music, lectures, etc.), the sense of spatial *intimacy* will be affected by the interval between the time the direct sound reaches the listener and the time that the first reflected sound arrives. There are indications that particularly for music, the optimum condition occurs when the time delay is 0.02 sec. The sense of intimacy declines, then, as the interval exceeds this optimum.

Reverberation

Reverberating signals tend to mix with other direct and reflected sounds in an enclosed space, and this may affect the blending or clarity of various tones. So the problem of sound persistence is a basic spatial consideration in the development of an appropriate background for sonic communication —particularly in larger spaces.

The objective is to provide conditions for full quality sound that is compatible with the dominant activity, without excessive hardness or muddling.

Conditions for Speech. In *live* spaces where reverberation times are longer than 2.0 sec, the intelligibility of speech becomes increasingly more difficult because of a tendency for signals to become muddled and confused by superimposed reflections. Clarity of speech steadily improves below 2.0 sec; approaching an optimum condition at about 1.0 sec.

No further improvement is observed below 1.0 sec. However, there may be some improvement in signal intelligibility with lower reverberation times because of the concurrent dampening of interfering background noise intensities.

Table 1-2.3 Reverberation Time

	Reverberation Time
Optimum for speech Generally too *dead* for music	Below 1.0 sec
Good for speech Fair for music	1.0–1.5 sec
Fair for speech Good for music	1.5–2.0 sec
Poor for speech Fair to poor for most music Good for liturgical music	2.0 + sec

Conditions for Music.   Hearing conditions for music are more a matter of tradition and taste than intelligibility. For this reason, there tends to be a wide range of acceptable reverberation times. But in general, the required ranges are summarized in Table 1–2.4.

Table 1-2.4 Typical Reverberation Time Criteria for Music

	Reverberation Time
Relatively small rehearsal rooms	0.8–1.0 sec
Chamber music	1.0–1.5 sec
Orchestral music, choral music, contemporary church music	1.5–2.0 sec
Large organ, liturgical choir	2.0 + sec

These intervals generally apply at 500 cps and above. Longer intervals are acceptable at lower frequencies (and are generally desirable for most music). Figure 1–2.6 indicates the nature of this variation.

FIGURE 1–2.6

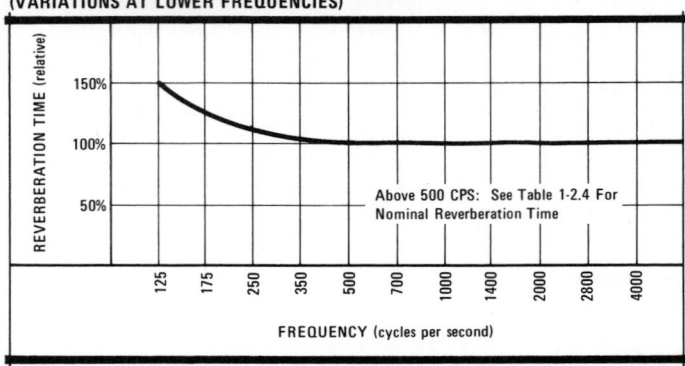

TYPICAL REVERBERATION TIME CRITERIA FOR MUSIC
(VARIATIONS AT LOWER FREQUENCIES)

FIGURE 1–2.7

BACKGROUND NOISE (SPEECH INTERFERENCE LEVELS)

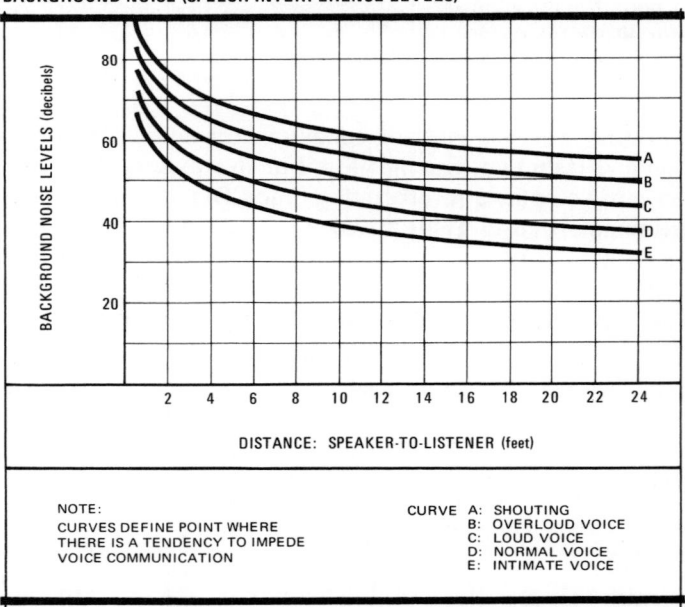

Background Spatial Context

Background noise in a space constitutes a context or reference level against which significant sonic signals must be compared for contrast and clarity. As a practical observation, it is generally the background noise and not the true physiological threshold of hearing that determines whether an individual will detect a meaningful sound. If the signal to background ratio is low, the marginal lower intensity sounds (such as consonants in speech) will tend to become unintelligible.

So, background noise at intensity levels that approximate or exceed minimum speech intensities will partially or decisively prevent comprehension of speech signals. When the signal intensity is known, the designer can expect a signal that exceeds the background intensity by 3 db to be barely perceptible; while a difference of 7 db or more is a clearly perceptible difference.

In order for verbal communication to continue under adverse background conditions, it is necessary for the individuals involved to increase the intensity of their speech (1) by moving closer together and thereby reducing the source to listener distance, and/or (2) by raising voice levels in an attempt to improve sonic contrast between the signal and the background. (See Figure 1–2.7 for rule of thumb relationships.)

Limitation of reflected noise

To some extent, internal background noise can be dampened by the addition of sound absorbing materials and finishes. Figure 1–2.8 indicates the approximate limits involved.

In general, spaces that contain operating machines or group interaction should be treated as absorbent interiors. In this way, the adverse influence of *reflected noise* is minimized; although *direct noise* in the vicinity of the source is unaffected by this procedure.

Activities which produce low noise levels require less absorption. These spaces may actually utilize hard reflective surfaces to facilitate effective distribution of reflected speech signals over or around a group of people. This need for hard reflective surfaces becomes particularly critical if speech must be projected over distances greater than 25–30 ft.

Limitation of direct noise

When evaluating direct noise in a space, two contributing factors can be manipulated by the designer: (1) the noise level of the equipment that must function within the space itself, and (2) the external sound transmission through walls, ceilings, and floors.

In this regard, optimum criteria are somewhat difficult to formulate as a precise numerical expression. This is because of the previously discussed need to express sonic criteria as a spectrum-type relationship between sound intensity and frequency.

When this is done, it is noted that, in general, the human ear can

FIGURE 1–2.8

BACKGROUND NOISE (ROOM SOUND ABSORPTION)

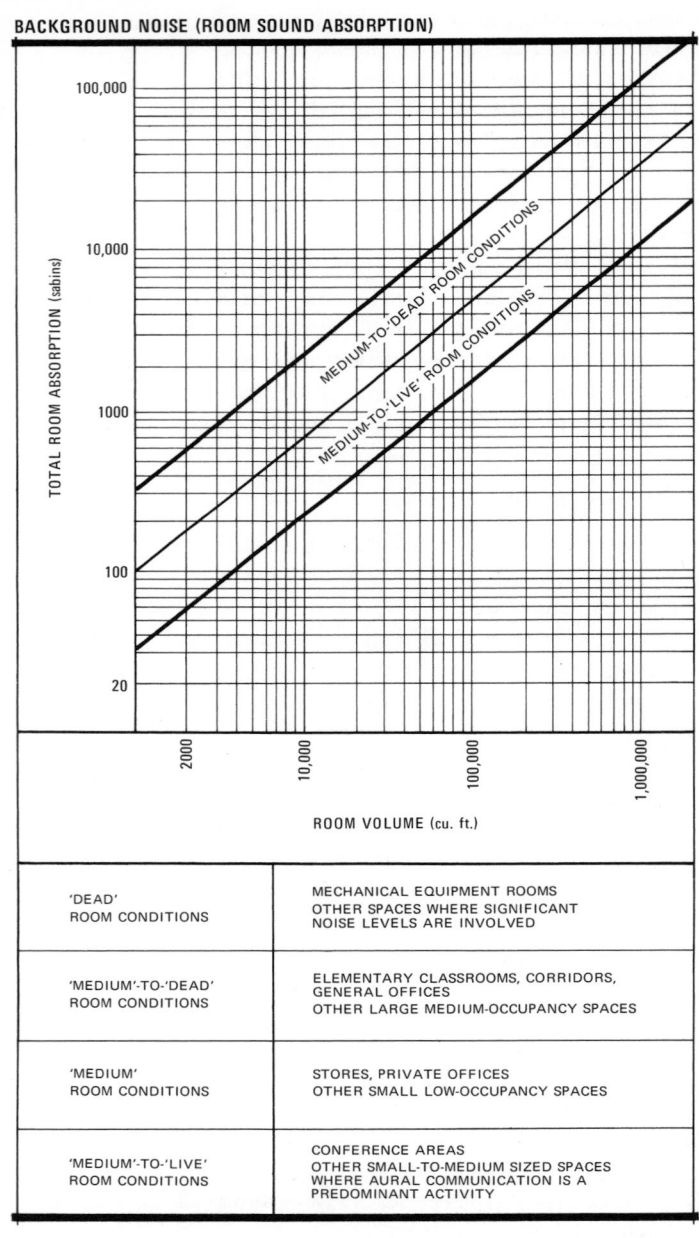

'DEAD' ROOM CONDITIONS	MECHANICAL EQUIPMENT ROOMS OTHER SPACES WHERE SIGNIFICANT NOISE LEVELS ARE INVOLVED
'MEDIUM'-TO-'DEAD' ROOM CONDITIONS	ELEMENTARY CLASSROOMS, CORRIDORS, GENERAL OFFICES OTHER LARGE MEDIUM-OCCUPANCY SPACES
'MEDIUM' ROOM CONDITIONS	STORES, PRIVATE OFFICES OTHER SMALL LOW-OCCUPANCY SPACES
'MEDIUM'-TO-'LIVE' ROOM CONDITIONS	CONFERENCE AREAS OTHER SMALL-TO-MEDIUM SIZED SPACES WHERE AURAL COMMUNICATION IS A PREDOMINANT ACTIVITY

tolerate higher levels of low frequency noise; while high frequency noise must be held to lower intensities. This tolerance is true both from the standpoint of annoyance and ability to understand speech signals.

Noise Criteria Curves (Beranek) express this tolerance as a series of spectral background curves (see Figure 1–2.9). These curves represent a performance standard for identifying permissable background sound levels for each of the eight major octave bands. The objective is to provide a means through which a general noise level for unoccupied (but otherwise

FIGURE 1–2.9

FAMILY OF NOISE CRITERIA CURVES

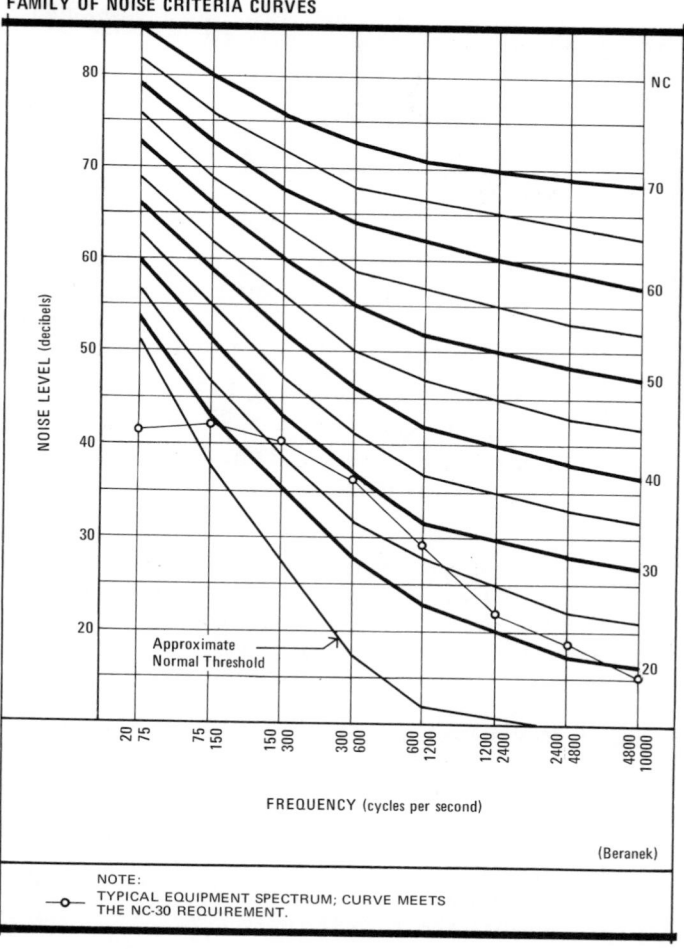

FREQUENCY (cycles per second)

(Beranek)

NOTE:
TYPICAL EQUIPMENT SPECTRUM; CURVE MEETS THE NC-30 REQUIREMENT.

normally operating) rooms can be specified and measured in a manner that reflects both intensity and frequency (see Tables 1–2.5 and 1–2.6).

These background noise limitations should influence the selection of all *noise-generating equipment* (machines, air conditioning components, fluorescent ballasts, etc.) and all *noise-abating elements* (sound attenuators, partitions, etc.). For any of these noise influences, an intensity spectrum can be developed and related to the Noise Criteria Curves. This makes it possible to estimate whether or not the source in question can be expected to become an adverse environmental influence within a given space-activity context. It is intended that for a specified criteria level, no frequency on the applicable curve should be exceeded.

Table 1-2.5 Background Noise (Typical Criteria)

Background Noise Levels	Sonic Conditions (Re: Figure 1-2.9)	Suitable Use
To NC-25	Very Quiet	Restful Contemplative
NC-25–35	Quiet Normal conversation: 10–30 ft	Reception Discussion Classroom Office General home
NC-35–45	Moderately Noisy Normal conversation: 6–12 ft Raised voice: 10–30 ft	General office (limited discussion)
NC-45–50	Noisy Normal conversation: 3–6 ft Raised voice: 6–12 ft Telephone use becomes slightly difficult	General office (w/o machines) Drafting
NC-50–55	Noisy Normal conversation: 1–2 ft Raised voice: 3–6 ft Telephone use becomes difficult	Typing Clerical (w/machines)
NC-55–70	Very Noisy Raised voice: 1–2 ft Telephone use becomes unintelligible at higher levels	Industrial

Table 1-2.6 Noise Criteria Curves (Typical Recommendations)

	Applicable Noise Criteria Curve (*Re: Figure 1-2.9*)
Concert halls, broadcast studios	NC-15–30
Theaters (w/o amplification)	NC-20–25
Music rooms, classrooms	NC-25
Apartments, assembly halls, motel rooms	NC-25–30
Offices	NC-25–35
Hospitals, churches, libraries, movie theaters	NC-30
Heavy circulation areas	NC-45
Assembly areas (w/amplification)	NC-50

Performance of isolating barriers and enclosures

Sound can be transmitted from an adjacent or remote *source space* by two basic methods. *Structural transmission* involves actual flexural vibration of the intervening partitions; while *airborne transmission* involves direct passage of sound through openings and pores. These are essentially parallel paths, and the intensity of the transmission will depend on which of the two paths offers the least resistance.

The difference in intensity between the emitted sound in the *source space* and the perceived sound in the *receiving space* is measured in decibels and is termed *noise reduction*. For example, a 30 db reduction is a transmission coefficient of 0.001 (i.e., 1/1000 of the source intensity passes through the barrier to the receiving space); a 40 db reduction is a transmission coefficient of 0.0001. (See Table 2-2.2.)

As previously noted, however, any meaningful analysis of sonic factors must evaluate the full spectrum of intensities and frequencies.

Sound transmission *Sound Transmission Class* (STC) is a useful device for preliminary evaluation and prediction of noise transmission through a barrier. This method relates to common sounds, such as those emitted in human speech, in music, by a barking dog, etc. Airborne energy from these signals will induce subtle vibrations in a wall, ceiling, or floor. This action, in turn, induces sound in the adjacent space. STC, then, is a method for specifying the performance of a barrier in resisting the transmission of these noises. (Also see discussion of sound transmission in Chapter 2-2.)

Prespecified spectral contours are defined in Figure 1–2.10. For a given

FIGURE 1–2.10

FAMILY OF 'STC' CURVES

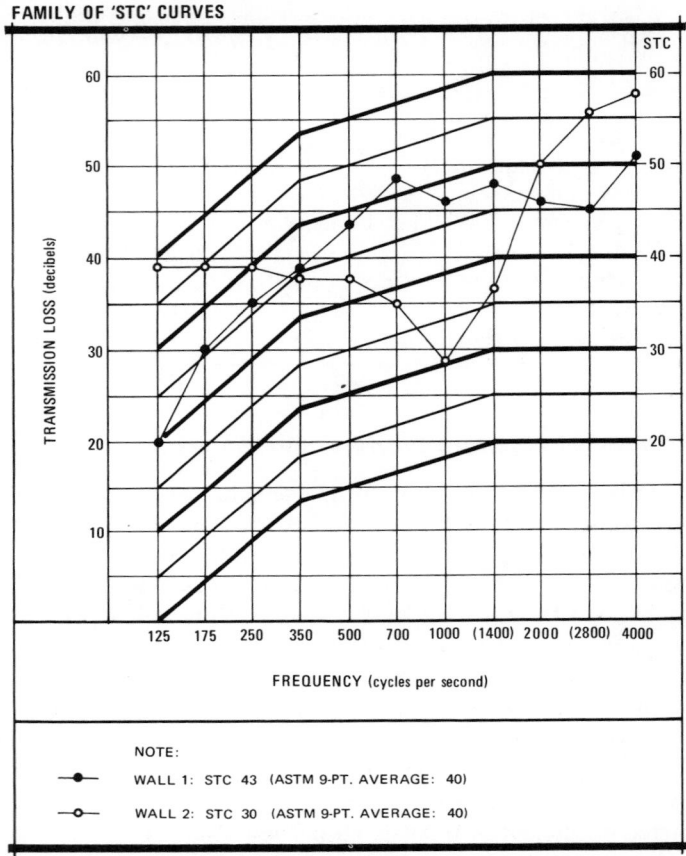

NOTE:
— ● — WALL 1: STC 43 (ASTM 9-PT. AVERAGE: 40)
— ○ — WALL 2: STC 30 (ASTM 9-PT. AVERAGE: 40)

barrier, the measured noise reduction values at each frequency band must lie on or above the *middle section* of the relevant curve. The measured values may fall below the two *end sections* by up to 3 db, provided the average deficiency in these extreme frequencies is less than 1 db.

Impact noise   Impact noise is sound generated by an object striking, vibrating, or sliding against a component of the building. This includes the effect of footsteps, moving furniture, dropped objects, doors slamming, mechanical equipment vibration, etc. Such actions will set the affected barrier or enclosure in vibration; and this, in turn, induces an impact sound on both sides of the barrier.

Table 1-2.7 Typical Transmission Loss Criteria

Typical Applications	Hearing Conditions	Transmission Loss	
		NC-25* Background	NC-35* Background
Privacy not required. Partition used only as a space divider.	Normal speech can be understood easily and distinctly through the wall.	STC 35 (or less)	STC 30 (or less)
Suitable for dividing noncritical areas. Provides fair degree of freedom from distraction.	Loud speech is understood fairly well. Normal speech can be heard, but is not easily understood.	STC 35–40	STC 30–35
Provides good degree of freedom from distraction. Suitable for junior executives, engineers, etc.	Loud speech can be heard, but is not easily intelligible. Normal speech can be heard only faintly (if at all).	STC 40–45	STC 35–40
Provides a confidential degree of speech privacy. Suitable for doctors, lawyers, senior executives, apartments, etc.	Loud speech can be heard faintly, but not understood. Normal speech is inaudible.	STC 45–40	STC 40–45
Suitable for dividing private offices from noisy areas that contain typewriters, computers, etc.	Very loud sounds, such as loud singing, brass musical instruments, or a radio at full volume can be heard only faintly.	STC 50 (or more)	STC 45 (or more)

* See discussion of "Noise Criteria Curves" (NC).

FIGURE 1–2.11

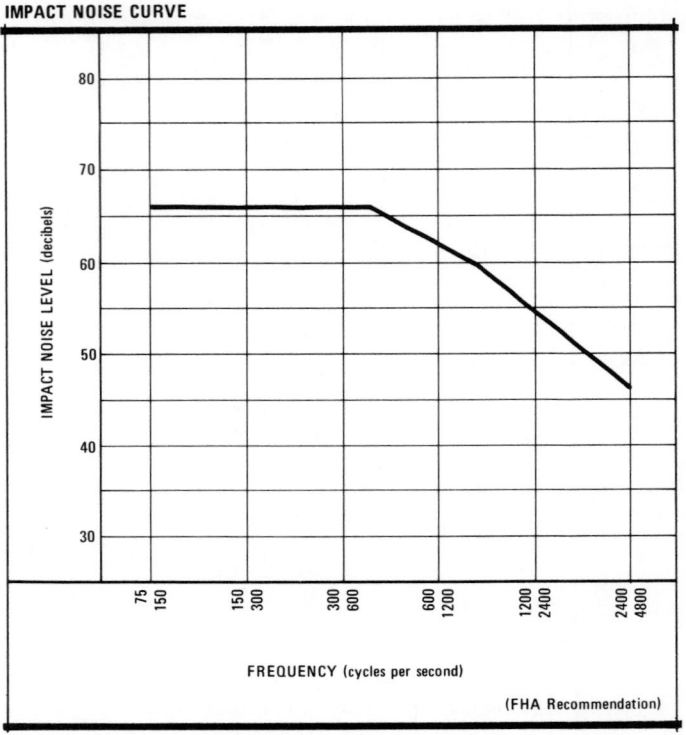

IMPACT NOISE CURVE

Impact Noise Ratings (INR) are one useful device for simple evaluation and prediction of the ability of a given floor construction to resist (and therefore reduce) more common impact sounds. Essentially, this rating is

Table 1-2.8 Typical Impact Noise Criteria

	Low Background Noise Level	*High Background Noise Level*
Corridor floors	+ 5	+ 2
Apartment floors	0	− 2
Floor above noisy public space	− 5	− 8
Floor above critical quiet space	+ 5 (or better)	+ 5 (or better)

NOTE: Numbers refer to the number of decibels by which the construction exceeds or fails
to meet the standard curve (see Figure 1-2.11).

the number of decibels that the curve of measured impact transmission must be shifted in order to bring it within proximity of a standard noise curve (see Figure 1–2.11).

A +*INR* (i.e., the measured values lie *below* the standard curve) indicates a tendency toward superior performance. A −*INR* (i.e., the measured values lie *above* the standard curve) indicates a tendency toward inferior performance. (Also see discussion of impact noise in Chapter 2–2.)

The influence of supplementary noise screens and masks

While previous discussions have dealt primarily with the negative aspects of background noise in reducing the sonic contrast of signals intended for communication, background noise also affects the psychological sense of *privacy*. For example, an office in a rural or residential area may be classified as *noisy* by people who are accustomed to the higher levels of background noise commonly found in downtown locations. The description *noisy* refers, in this instance, to distraction or lack of sonic privacy. It refers to the fact that in an extremely quiet space, low intensity sounds and irrelevant distant conversations may be easily heard; and this becomes distracting.

Ambient noise can now serve a positive function by establishing an artificial threshold that will mask these remote or low intensity distractions. Such a *noise screen* will support a sense of sonic privacy in a large open space, and it can also supplement the attenuation value of partitions.

FIGURE 1–2.12

TYPICAL EFFECT OF BACKGROUND NOISE SCREENS

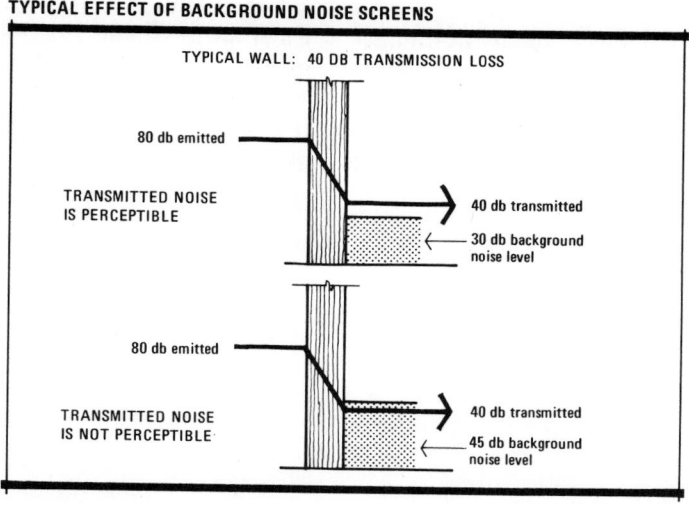

Figure 1–2.12 suggests how a partition that is inherently only moderately effective as a barrier to sound transmission may be completely suitable when a continuous background noise (produced by ventilation fans, moderate traffic, etc.) becomes an innocuous part of the sonic environment. But when this ambient noise screen is reduced or removed, the artificial threshold lowers and the same partition may be inadequate for effective sonic isolation.

When intentionally designing such a sonic mask, applicable spectral characteristics are generally defined by the appropriate *Noise Criteria Curve* (see Figure 1–2.9 and Table 1–2.6).

Wide-band masks   As discussed previously in this chapter, a *white noise* is a full spectrum sonic distribution that is somewhat analogous to *white light*. It implies fairly uniform distribution of sound energy at all frequencies. (A typical white noise is air escaping from a jet.)

In general, an effective mask or noise screen is a featureless white noise that produces slightly higher intensities at the low frequencies, and lower intensities at high frequencies. Some adjustability in intensity is desirable, and extreme care should be taken to insure that the mask does not reach sufficient intensity that it conflicts with normal communication of speech or music. Use the appropriate Noise Criteria Curve for guidance.

Pure tones and intermittent noise backgrounds   Intermittent intensities and pure tones are generally not appropriate as an effective mask or screen because they may, in themselves, introduce an adverse influence into the environment. High frequencies are particularly troublesome in this regard.

However, if either *one* of these must be present in an environment intended for sonic communication, it is generally desirable to reduce the source intensity so that the critical frequency falls 5–10 db below the applicable Noise Criteria Curve (see Table 1–2.6). If the noise is *both* intermittent and a pure tone, the acceptable background level should fall 10–20 db below the usually applicable curve.

The Thermal Environment

HEAT AND ATMOSPHERIC VARIABLES

Factors involved in the thermal environment are not directly related to communication or orientation. Unlike some of the more subtle considerations related to perception of visual or sonic detail, then, thermal factors exert a relatively minor influence on individual participation and performance—*unless* these factors become adverse to the extent that they induce physiological stress.

The Human Body

The human body receives chemical energy from food. This energy is, in turn, transformed into other forms of energy by the body; and heat is released by this *metabolic process.*

Metabolic rate (i.e., the rate of heat emission) will vary significantly with movement or activity. To a less significant degree, it will vary with the size of the individual, with age, and with the sex of the individual.

For a given person who is steadily involved in the same activity, however, it appears that his metabolic rate will be stable in the environmental temperature zone of approximately 70–90°F. Below this range, the adaptive effect of shivering may begin to increase the metabolic rate. Above the stable range, the metabolic rate may again tend to increase; not as an adaptive procedure, but because the internal chemical action will increase as the deep body temperature moves higher.

The heat flow interaction between the body and the surrounding environment takes place through convection, conduction, radiation, and evaporation. In this sense, the direction of heat flow may be either to or from the body. If the algebraic sum of heat flow quantities is either (1) a net flow to the body, or (2) a net flow from the body at a rate below the metabolic rate, this implies a heat storage condition within the body that will tend to produce an increase in the temperature of the deep body tissues. Similarly, a net flow from the body at a rate that exceeds the metabolic rate will produce a decrease in the deep body temperatures. If either of these actions is continued over even a moderate period of time, adverse physiological effects result (including death).

Methods of body heat adjustment

When in a relaxed state, a human occupant will generate approximately 400 btu/hr. Figure 1–3.1 indicates the general method by which this body heat is removed and dissipated.

FIGURE 1–3.1

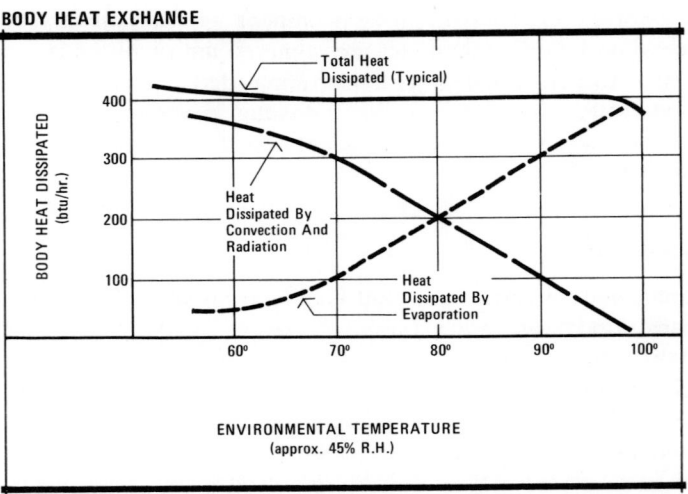

The dissipation curve begins to drop sharply as the environmental temperature approaches and exceeds the deep body temperature of 98.6°F. This indicates the diminishing capability of the body to cool itself effectively in these environmental circumstances. The result of this condition is eventual heat exhaustion for the individual.

In the range of environmental temperatures below approximately 70°F, additional body heat is indicated in the diagram. This reflects the adaptive effect of involuntary physical activity (such as shivering). As this becomes more pronounced and excessively prolonged, the individual can begin to suffer from exhaustion and overcooling of the skin (exposure).

Adjustment to warm environments When the environmental air temperature lies within the neutral zone for relaxed human occupancy (typically the low to mid-70° range), the flow of blood through the body utilizes the deep-seated arteries and returns through the deep veins. As a result, relatively little heat is transferred from the blood to the surface skin.

As the environmental temperature begins to rise above the neutral zone, the superficial veins (particularly those in the limbs and extremities) begin to dilate. The diversion of returning blood to these superficial veins causes the blood to act as a heat-convecting medium; increasing the flow of heat from the deeper tissue to the surface and causing the skin temperature to rise. This action increases the loss of heat at the skin by radiation and convection to the external environment.

When dilation of the veins reaches the normal maximum, skin temperature is somewhat uniform over the entire body and has reached a normal maximum. The body is therefore incapable of increasing body heat dissipation by convection and radiation. When environmental temperatures continue to rise, therefore, sensible sweating becomes a mechanism to increase body heat dissipation through evaporative cooling of the skin.

The rate of sweating will vary with the heat dissipation requirements. At approximately 98.6°F, where convection and radiant losses are negligible because there is no essential temperature difference between the human circulation system and the surrounding environment, evaporation tends to become the exclusive body regulative process.

However, the sweating mechanism tends to fatigue with time. So the rate of sweating will tend to decrease after several hours of exposure to adverse conditions. This fatigue is partially related to the fact that the sweating process consumes body fluids. (Approximately one pound of body fluid is consumed for each 1000 btu dissipated; an action that produces losses in body weight.)

Adjustment to cold environments When environmental temperatures drop, the previously discussed processes are, of course, reversed. When temperature falls into the neutral zone, then, the pores tend to close, so relatively little heat is dissipated through evaporation. Most of the body heat is being dissipated by convection and radiation; with experiments indicating that radiation losses constitute the most important single means of dissipation in this environmental range.

As temperatures fall below the neutral zone, further physiological changes occur as the body attempts to minimize heat losses due to radiation and convection from the skin surface. The superficial veins become constricted, while the deeper veins dilate. Body heat is therefore retained within the deeper tissues, while the skin surfaces (particularly the extremities) are permitted to cool. This drop in extremity surface temperature is sometimes quite significant; a condition which, of course, tends to decrease the loss of heat by radiation and convection.

As environmental temperatures continue to drop and the variation in blood flow can no longer compensate adequately, deep tissue temperatures drop and involuntary physical activity (shivering) becomes a means to generate additional body heat. This action does not fully compensate for the drop in deep tissue temperature, but rather attempts to prevent a continued drop.

Acclimatization   As environmental temperatures rise or fall and stabilize at new levels (as may occur with seasonal changes), the more permanent constriction or dilation of the superficial veins requires the body to respond by varying the total volume of blood. This variation may approximate as much as 20%, with the greater volume required for the warmer environmental conditions.

Such lasting adjustments in blood supply will require approximately four days or more to accomplish, and this is a major factor in determining the general *acclimatization* of the body to warm or cool climatic conditions.

Acclimatization will also affect sweating ability. With continued exposure to overheated conditions, the maximum sweating rate will tend to increase over a period of several weeks.

Variations in activity level   The previous discussions have described the action of a typical human occupant in a relaxed state. In this condition, body heat is generated at approximately 400 btu/hr.

Table 1–3.1 shows how metabolic rate increases with increasing activity. When the occupant is participating in more strenuous activities, then, the rate of body heat dissipation must be increased dramatically.

This extremely variable rate of body heat generation suggests that identification of a *neutral zone* in the environment varies with the nature of the activity being performed. An environmental temperature of 65°F may be excessively warm for an active group of men doing heavy manual work; but the same condition may be quite cool for a group of women involved in more leisurely activity.

The table also includes an attempt to approximate neutral comfort ranges for the various activity classifications. In each case, the temperatures shown assume that the participating individuals will remain involved for a moderate period of time.

Table 1-3.1 Typical Effects of Activity Variations (Thermal)

Activity	Body Heat Generation (Approximate)	Typical Neutral Zone* (Environmental) Air Temperature)
Heavy exercise	1500–2500 btu/hr	55°–60°
Moderate manual work	750–1500 btu/hr	60°–65°
Normal circulation	600–750 btu/hr	65°–70°
Normal rest—seated (see Figure 1-3.1)	400 btu/hr	70°–75°
Normal rest—bed	250 btu/hr	75°

*Applies for typical individuals who will tend to remain under these activity conditions for at least 2–3 hr.

Skin temperature

Skin temperatures vary considerably. In the normal range of environmental temperatures, the *average* skin temperature of the occupant at rest may vary from a maximum of approximately 97° to a minimum of 86° without any change in deep tissue temperature. The skin temperature of the extremities may drop even lower under normal environmental conditions.

As the environment warms above optimum, then, an early effect is that both trunk and extremity skin temperatures rise toward the maximum values. (This action increases the dissipation of heat through convection and radiation.) Further warming causes sensible sweating.

Optimum comfort for clothed individuals at rest appears to occur when the average skin temperature is in the range of 91–93°F. For active individuals, this optimum value is lower, sometimes falling as low as 85°F when heavy exercise is involved.

The sensation of temperature change In addition to affects associated with metabolic processes, the body also reacts to external hot or cold stimuli of a spatial or localized nature. This sensation of warmth or coolness depends on changes in skin temperature.

However, the body is not a good *thermometer* for estimating the intensity of temperature variations. The general stabilizing effects of acclimatization may affect such judgments. Furthermore, the skin is more sensitive to heating than to cooling. Some research groups have reported that a warming sensation is produced by a rate of increase in skin temperature of 0.002°F/sec or more; while a cooling sensation requires a decrease in skin temperature of at least 0.007°F/sec.

Response to humidity

Dissipation of body heat through evaporation is directly proportional to the quantity of perspiration that is evaporated. Since the evaporation rate will vary significantly with the saturation of the surrounding air, then, relative humidity is a particularly critical consideration which affects the ability of the body to dissipate heat when environmental temperatures are high.

As a corollary, relative humidity will affect the highest range of environmental temperatures to which the body can be exposed (in still air) and still maintain effective heat regulation through evaporation. One study has reported typical limits as follows:

$$88°F \quad \text{at } 100\% \text{ RH}$$
$$99.5°F \text{ at } 51\% \text{ RH}$$
$$113°F \quad \text{at } 18\% \text{ RH}$$
$$126°F \quad \text{at } 0\% \text{ RH}$$

As environmental temperatures cool, however, the influence of humidity on the ability of the body to dissipate heat becomes less significant. When the pores are closed and sweating action is no longer involved, humidity only affects heat that is dissipated through breathing. This heat transfer depends on the taking in of relatively unsaturated air, with subsequent expelling of saturated air. The greater the difference in saturation, the greater the body heat loss; so this heat transfer is greatest when environmental humidity is low. However, in most cases, heat dissipation through this means is less significant than other methods previously discussed.

The Thermal Environment

In summary the human body functions as a heating and cooling system. It can, within limits, effectively and instinctively adjust to environmental conditions that vary from optimum.

Essentially, the body functions to lose heat at a controlled rate to moderately cooler air and cooler surfaces. However, if this cooling action is either too slight or excessive, the result can be occupant discomfort and physiological stress.

Then too, the thermal environment is much more complex than simple temperature relationships. The occupant also emits heat through exhalation and through evaporation of perspiration. As a result, the body responds to the quality of air in the space, both in terms of *freshness* and humidity. So ventilation, filtering, and water vapor control become additional criteria which must be related to the heat transfer potential of convection and radiation.

Thermal comfort, then, is a near-optimum condition of equilibrium; an environmental background that facilitates favorable atmospheric interchanges, maintains a favorable skin temperature, and minimizes the need for any of the more extreme physiological adjustments that are required to maintain a balance between internal body-heat gains and external heat losses.

PERCEPTION OF THE THERMAL-ATMOSPHERIC BACKGROUND

When the designer defines the performance characteristics of the thermal environment, he generally strives to achieve and maintain a somewhat neutral background condition. The initial objective is to minimize adverse thermal and atmospheric influences that could otherwise impede either participation in an activity or effective performance of a task.

In the sense of system management, this requires manipulation of several related environmental factors: (1) manipulation of the temperature and humidity of the air mass that surrounds the body; (2) manipulation of the movement and composition of this air mass; and (3) manipulation of the temperature of major surfaces that surround the body.

Environmental Air Temperature and Humidity

The thermal variables of environmental air temperature and humidity have been combined into a single empirical index called *effective temperature*. This relationship permits the designer to identify and anticipate an approximate comfort zone.

As shown in Figure 1–3.2, the effective temperature relationship applies: (1) for reasonably still air conditions (15–25 ft/min); (2) when the occupants are seated at rest or doing very light work; and (3) when the room surfaces are at or near the environmental air temperature.

The comfort zone

The *winter* optimum for normally-clothed people appears to be near 67–68 ET (effective temperature). In northern climates, 20–30% relative humidity is an approximate compromise between the excessively low range that will cause drying and dehydration, and the higher humidities

FIGURE 1–3.2

EFFECTIVE TEMPERATURE CHART

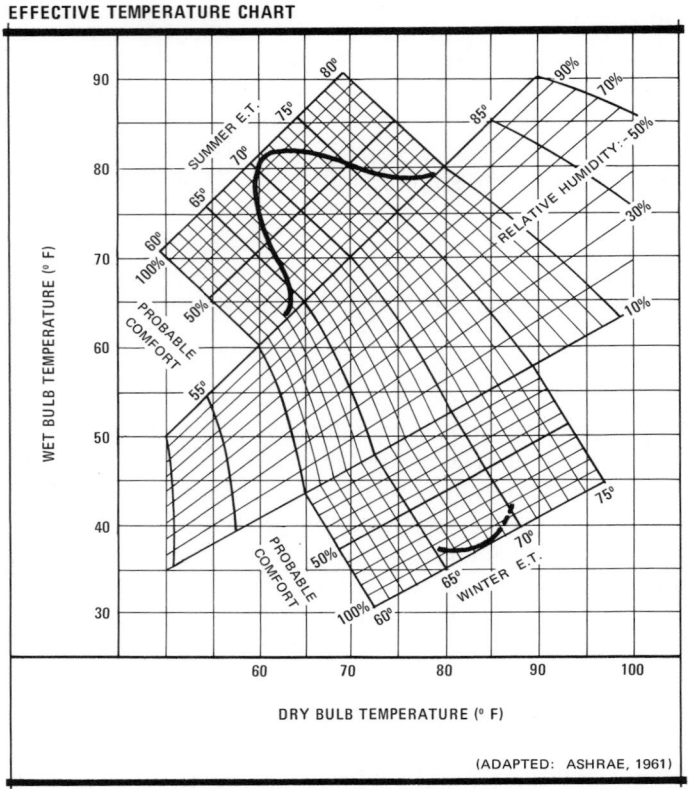

(ADAPTED: ASHRAE, 1961)

that will cause excessive condensation during cold outdoor periods. In this sense, the optimum effective temperature range is achieved with an environmental air temperature of 74°F (dry bulb) and 30% RH, with relatively still air moving at a rate of 15–25 ft/min.

The *summer* optimum for normally-clothed people appears to be near 71 ET (effective temperature). To achieve this, a common target is 76°F (dry bulb) and 50% RH, when air is moving at a rate of 15–25 ft/min.

As a general evaluation, then, optimum environmental air temperatures (for occupants at rest) generally range between 73–78°F, at the same time that relative humidity ranges between 25–60%. Note that the humidity tolerance within the comfort zone is much greater than the temperature tolerance. This indicates that temperature must be controlled much more precisely.

Variations from the assumed conditions

Variations from the general conclusions indicated in Figure 1–3.2 are as follows:

Densely occupied spaces   For congested occupancy conditions, comfort will occur at a somewhat lower effective temperature than that indicated in the diagram. This change is due to the effect of radiant heat transfers between occupants.

Activity   Occupants who are involved in more strenuous activities will also require lower effective temperatures than those indicated in the diagram. (See previous discussion of "Variations in Activity Level.)

Duration of occupancy   The diagram is based on occupancy durations of three hours or more. For short-term occupancy, effective temperatures should generally be maintained at levels somewhat higher than those shown.

Sex of the occupant   There are some indications that women tend to prefer effective temperatures slightly higher than those preferred by men. This is due to differences in metabolism.

Age of the occupant   There are similar indications that men and women over forty-years tend to prefer effective temperatures slightly higher than those preferred by younger people.

Geographical variations   The diagram is based on subjective conditions typically found at 42° north latitude. When estimating the probable comfort zone, allow 1 ET increase in effective temperature for each 5° reduction in latitude. (See previous discussion of "acclimatization")

Condensation

Condensation is a factor in any evaluation of temperature and humidity. This condition will occur when the temperature of any surface falls below the dew point of the ambient air. For a constant interior humidity and air temperature, the point at which condensation will occur on or within an assembly varies *directly* with the inducing temperature influence and *inversely* with the heat transmission characteristics of the materials. The better the insulation value, the more remote and intermittent is the problem of condensation.

For this reason, thin metal and glass (both of which conduct heat quite readily) are most susceptible to condensation. So cold air ducts, metal air supply diffusers, cold water pipes, and window sashes are particularly vulnerable to this problem; as is the window glass itself.

Table 1–3.2 indicates typical condensation conditions for various exterior wall materials. The use of venetian blinds, draperies, and other window coverings may alter this to a slight degree.

When condensation conditions are likely, the problem can be controlled in two ways:

(1) By reducing the relative humidity of the interior air mass. This action will lower the temperature range at which the problem will occur. But it may also reduce occupant comfort and well-being if the humidity is maintained too low. For this reason, it is generally desirable to maintain relative humidity at a level of 20–30% or higher.

(2) By increasing the insulation value of the surface. This will also lower the temperature range at which the problem will occur. This is accomplished with an air space (double or triple glazing) or by otherwise insulating the surface involved (such as the encasement of cold ducts and cold water pipes; the utilization of less conductive materials in place of metal; etc.).

Table 1-3.2 Condensation

| | Approximate Outside Temperature at Which Condensation Will Begin to Occur on the Inside Face | | | |
| | Relative Humidity (70° interior air) | | | |
	20%	30%	40%	50%
Aluminum or steel	16°	28°	38°	46°
Glass: single glazing	13°	25°	36°	45°
Glass: double glazing	—	−20°	1°	20°
8-in face brick wall	—	−17°	3°	22°
6-in concrete roof	—	−26°	−3°	17°

Air Distribution and Composition

As discussed previously in this chapter, a human occupant will emit heat to the surrounding air. This action will, in turn, raise the temperature of this air. Furthermore, evaporation of perspiration and support of the breathing process will modify the composition of the air mass.

Machines and similar heat-producing devices will also affect the air mass, as does the transmission action of various building surfaces. While air movement is not the only means of thermal control, then, the air mass is particularly significant in the sense that it affects many aspects of the thermal-atmospheric environment.

Compensating effects of air motion

When air temperature and humidity exceed the human comfort range, increased air motion can become a compensating influence and return the sensory environment toward a condition of thermal equilibrium. In this sense, increased air velocity tends to raise the comfort zone defined in Figure 1–3.2.

Table 1–3.3 describes how the human occupant will perceive various levels of air motion. Table 1–3.4 indicates a number of subjectively com-

Table 1-3.3 Background Air Motion (Typical Behavioral Patterns)

Air Speed at Head Level	Subjective Evaluation
less than 15 ft./min	Complaints about stagnant air when other atmospheric conditions lie in the comfort range
15–25 ft/min	Favorable; basis for *effective temperature chart* (Figure 1-3.2)
25–50 ft/min	Favorable conditions when atmospheric conditions lie in the comfort range Summer comfort range: 71–73 ET
50–100 ft/min	Subtle awareness of air movement, but generally comfortable when temperature of the moving air is at or slightly above room air temperature Summer comfort range: 73–75 ET
100–200 ft/min[1]	Constant awareness of air movement, but generally pleasant Summer comfort range: 75–77 ET
200–700 ft/min[1]	Increasingly drafty conditions; increasing complaints about the adverse effects of *wind* in disrupting a task, an activity, or personal composure

[1]There are some preliminary indications that localized and intermittent introduction of moderate velocity room temperature air at head level will induce temporary stimulation in an occupant who is involved in long-term activities or work tasks.

parable temperature, humidity, and velocity combinations—each of which combine to produce a somewhat similar effective temperature of approximately 71 ET.

Table 1-3.4   Background Air Motion (Compensation for Rise in Air Temperature and Humidity)

Environmental Conditions	Air Velocity Required to Approximate Near-Optimum Body Heat Transfer Conditions for an Occupant at Rest
76° 50% RH	15–25 ft/min
80° 35% RH	100 ft/min
75° 80% RH	100 ft/min
82° 30% RH	200 ft/min
75° 100% RH	200 ft/min
85° 35% RH	700 ft/min
79° 100% RH	700 ft/min

Air sanitation

Control of atmospheric contaminants is an important aspect of air handling. This affects comfort, and in some cases the health of the occupants.

Essentially, this aspect can be divided into three categories: (1) composition of air in terms of oxygen and carbon dioxide; (2) odors; and (3) dust control. The first two are controlled primarily through ventilation (i.e., the introduction of fresh outdoor air). The third category is primarily controlled with filters.

Composition of air   As a normal body process, oxygen is taken in and consumed; carbon dioxide and water vapor are expelled. An individual at rest may take in 16.2 cu ft of air per hour—consuming 0.504 cu ft of oxygen and producing 0.42 cu ft of carbon dioxide.

As the activity becomes more strenuous, the required quantities of air increase dramatically. For example, an individual walking at a rate of 4 mph will take in about 78.6 cu ft of air per hour—consuming 3.38 cu ft of oxygen and producing 3.0 cu ft of carbon dioxide.

High occupant densities or a prolonged period of occupancy can therefore begin to alter the composition of air in a sealed space (such as a mine, a submarine, an aircraft, a tightly sealed building, etc.). When this occurs, the quantity of oxygen is reduced, while the carbon dioxide increases. Table 1–3.5 shows the relationships that define both normal and abnormal conditions.

Table 1-3.5 Air Composition (Tolerance)

| | *Per Cent by Volume* | |
	Carbon Dioxide	*Oxygen*
Normal air condition	0.3%	20.6%
Objectionable condition	2.0%	16.0%
Dangerous condition	5.0%	12.0%

NOTE: 1.5% CO_2 is considered permissable for submarines. Approx. 3.5 cfm of outdoor ventilation air per occupant will usually maintain the composition at near normal conditions.

Table 1-3.6 Ventilation Rates (Typical Minimums)

	Air Space Per Person	*Outdoor Air Required*
Heating Season		
Air not conditioned		
Seated adults	100 cu ft	25 cfm/person
	200	16
	300	12
	500	7
Active adults	200 cu ft	145% of that required for seated adults
School children (average socioeconomic status)	100 cu ft	29 cfm/person
	200	21
	300	17
	500	11
School children (lower socio-economic status)	200 cu ft	180% of that required for average group
Heating Season		
Conditioned air* (air circulation: 30 cfm/person)		
Seated adults	200 cu ft	12 cfm/person
Cooling Season		
Conditioned air* (air circulation: 30 cfm/person)		
Seated adults	200 cu ft	4 cfm/person

*Assumes spray dehumidifier.   (Adapted: Ashrae, 1961)

Generally, a normal balance is preserved by adequate ventilation. As a general rule of thumb, 1.0 cfm of ventilation air per occupant will preserve the oxygen balance; and 3.5 cfm of ventilation air per occupant will preserve a CO_2 content below 0.6%. (In most cases, however, effective control of body odors and smoking requires higher quantities of ventilation air. See Table 1–3.6.)

Ventilation for odor control   Industrial processes, smoking, and body processes are representative sources of environmental odors. Such odors tend to cling to absorbent materials unless the environment is regularly *cleansed* through the introduction of fresh outdoor air.

The rate at which this ventilation air must be introduced into the system depends on the character of the process involved. Or, in the general case of human occupancy, it depends: (1) on the volume of space per occupant; (2) on the intensity of the activity; and (3) on the general physical and sociological conditions involved. Table 1–3.6 indicates the nature of typical recommendations which reflect these variables.

Filters are generally not effective for odor control because filters will only intercept solid particles. However, activated carbon is effective, and this technique is generally used to eliminate odors in closed systems (such as those used in submarines, mines, aircraft, etc.).

Atmospheric impurities   Airborne particles can produce irritation of the skin, the eyes, the lungs, and other critical human organs. More moderate concentrations can produce a general sense of discomfort.

These dispersions of material in air take several forms; each primarily controlled with filters:

(1) *Dust* refers to mechanically-generated solid particles that are usually dry. Particles over 10 microns in size are primarily of concern in that they collect on machinery and flat surfaces, and generally create problems of maintenance and cleanliness. Particles under approximately 10 microns can be inhaled into the human system. Of this group, the larger particles are trapped in the primary respiratory tracts; while particles under 5 microns are considered to be the most damaging in the sense of depth of penetration.

Particles smaller than about 20 microns in size tend to be invisible to the eye. But as a mass, they may be seen by reflection when an intense, directional beam of light passes through.

(2) *Smoke* involves solid particles of organic origin and is one of the finest forms of dust.

(3) *Fumes* involve the suspension of liquid or molten materials in free air. This is generally produced by chemical action.

(4) *Pollen* refers to moderate-sized organic particles (approximately 20–60 microns in size). These particles may be a source of irritation for some *allergic* individuals.

Mean Radiant Temperature

Body heat losses due to radiation depend primarily on the temperature of surrounding wall, floor, and ceiling surfaces. The more these surface temperatures vary from the average skin temperature, the greater is the radiant heat loss (or gain).

The *mean radiant temperature* (MRT) is a weighted average of the various radiant heat influences in the space. These surfaces and energy sources are weighted according to the area of the thermal influence that is exposed to (or *seen* by) the occupant's body surface.

$$\text{MRT} = \frac{A_1T_1 + A_2T_2 + A_3T_3 + \ldots}{A_1 + A_2 + A_3 + \ldots}$$

where: A_1 = the projected area of a specific surface or object
T_1 = the temperature of that surface

Note that locally variable thermal conditions may occur when the occupant is positioned near a large-area surface that is significantly warmer or cooler than the average.

In this regard, there are indications that floor surface temperatures are particularly critical and should not exceed 85°F in order to minimize possible discomforting effects on the skin temperature of the lower extremities. Similarly, to minimize possible adverse effects on the upper extremities, ceiling temperatures should be limited to 115°F or less.

Compensating effects of surface temperature

Although radiant heat transfer is not affected by air motion, there is a subjective relationship between effective temperature on one hand and mean radiant temperature on the other. For example, MRT can be used as a device to adjust the comfort zone; for as the cooling potential of environmental air is increased or decreased, this effect can be somewhat offset or cancelled by an *inverse* change in room surface temperatures.

In this regard, the effective temperature chart (Figure 1–3.2) is based on the condition that room surfaces are at or near the environmental air temperature (i.e., MRT is near the dry bulb temperature).

If room surface temperatures are *cooler* than this, comfort will normally occur at a higher effective temperature than is indicated in the diagram.

As an approximation, a 2°F drop in mean radiant temperature will be offset by a 1 ET rise in effective temperature.

Similarly, if room surface temperatures are *warmer* than the environmental air temperature, comfort will tend to occur at a lower effective temperature than is indicated in the diagram. Again, for estimating purposes, a 2°F rise in mean radiant temperature will offset a 1 ET drop in effective temperature.

Transitional Patterns

The experience of space is a dynamic one, with periodic or constant occupant movement between areas. When an occupant moves to an adjacent space, his orientation will change; and he may, at some point, become aware of the dominance of a new environment. This transition can therefore be developed to provide a sense of *continuity,* in the sense that luminous, thermal, and sonic influences are similar in the two adjacent spaces. Or, the transition can be developed to provide a sense of *environmental contrast* and change.

To carry this idea of environmental contrast a step further, we observe that when an individual moves from a space that is seriously deficient in light, heat, or sound into a space where these influences are noticeably greater in intensity, the new environment is generally considered to be more suitable and comfortable. But if, in successive steps, the various stimuli continue to increase in intensity, we find that at some point the positive influence wanes, and the same stimuli tend to become negative influences in the environment.

Figure 1–4.1 illustrates this relationship and shows that when each stimulus is evaluated as an influence on human experience and comfort, there is: (1) a *positive* range of experience in which the occupant is at least momentarily conscious of an improvement in his environment; (2) a *neutral* (comfort) range in which the occupant is free from the negative impingement of environmental deficiencies or excesses; and (3) a *negative* range of experience in which the occupant is at least momentarily conscious of a deterioration in his environment.

This diagram also indicates that when transient occupants are involved, these positive or negative interpretations will not be the same for individuals who are moving from different stages of adaptation. It is obviously important, then, for the designer to make some estimate of the individual's relative position in space—for subjective judgment of light, heat, or sound will be affected by the reference level established by the immediate previous environmental condition.

FIGURE 1–4.1

TRANSITIONAL ADAPTATION (TYPICAL BEHAVIORAL PATTERNS)

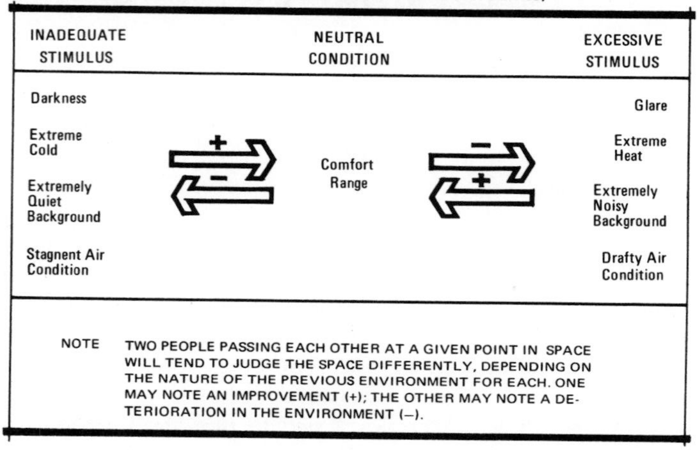

INADEQUATE STIMULUS	NEUTRAL CONDITION	EXCESSIVE STIMULUS
Darkness		Glare
Extreme Cold	Comfort Range	Extreme Heat
Extremely Quiet Background		Extremely Noisy Background
Stagnent Air Condition		Drafty Air Condition

NOTE   TWO PEOPLE PASSING EACH OTHER AT A GIVEN POINT IN SPACE WILL TEND TO JUDGE THE SPACE DIFFERENTLY, DEPENDING ON THE NATURE OF THE PREVIOUS ENVIRONMENT FOR EACH. ONE MAY NOTE AN IMPROVEMENT (+); THE OTHER MAY NOTE A DETERIORATION IN THE ENVIRONMENT (–).

ADAPTATION

Adaptation refers to the characteristic action of the human body which causes it to seek a state of equilibrium and enables it to physiologically adjust to the prevailing environmental conditions. In essence, after a short period of adjustment to a static environmental stimulus (light, heat, or sound), adaptation tends to neutralize the occupant's sensitivity to the intensity or magnitude of that stimulus. Once this adjustment has taken place, however, this state of individual adaptation may tend to intensify sensitivity to a new change.

When anticipating the effect of adjacent or successive environmental conditions, then, the designer must consider the probable temporary physiological condition of the senses: (1) the degree to which the iris of the eye is open or closed; (2) the degree to which various color receptors (cones) are fatigued or overly-sensitive; (3) the level of background noise to which the occupant has become accommodated; (4) the condition of the skin and the probable effect of changes in temperature, humidity, and air motion.

The occupant's subjective response to a new environmental condition will be affected by the imbalances that the new space creates for any of these senses—and the rate of change involved.

Abrupt Change or Environmental Contrast

Abrupt changes in the intensity of a given stimulus can exert a temporary but important influence on the attitude of the occupant. For example, a noticeable change in light, a significant change in air flow near the occupant, or an increase in background noise may each cause the occupant to become conscious of that particular sensory influence.

If the occupant is subjected to extremely low rates of information input, he may react with interest or a sense of relief from boredom. In this situation, then, the abrupt change is a useful and stimulating influence.

On the other hand, the occupant's reaction may also tend toward increased irritation and a feeling of tension when he is involved in a highly complex task or one that requires intensive vigilance. For this reason, when the occupant is subjected to more complex activities or tasks, environmental variations that affect adaptation should be avoided in favor of a more neutral condition.

Transitional influences related to temperature and air flow

Abrupt changes in environmental temperature or air flow in the vicinity of the body will both tend to cause an immediate increase or decrease in skin temperature. As a general rule, such transitional changes tend to be pleasant when they involve noticeable increases in air flow (above the 50–100 ft/min range, but below velocities that will be considered *windy*), with the moving air maintained at or slightly above the temperature of the principal air mass in the room.

When air velocity conditions are stable, there are preliminary indications that an environmental change must approximate 3 ET or 6° MRT in order for the change to be clearly perceptible. This conclusion appears to apply when the general thermal environment is near a condition of optimum comfort.

After a short period of exposure, the body will adjust its blood flow to compensate for the new condition, and the new environment tends to become the *norm*.

Transitional influences related to sound

A change of 3 db in background noise intensity will tend to be barely perceptible. The change will be clearly perceptible if it exceeds 7 db.

Changes in reverberation time are also perceptible if the occupant moves from a very soft, absorbent space to a hard space (or vice versa). Generally, the change in reverberation time between spaces must approximate 0.75 sec or more in order to be a significant subjective change.

(The effects of intermittent variations in white noise are discussed in Chapter 1–2.)

Transitional influences related to brightness

Visual transitions occur during movement between areas throughout the building. While spatial continuity is often appropriate and necessary, subtle contrast can become a useful source of stimulation and visual relief for an individual who moves about, but remains within the building for extended periods of time.

FIGURE 1–4.2

ADAPTATION AND SUBJECTIVE BRIGHTNESS

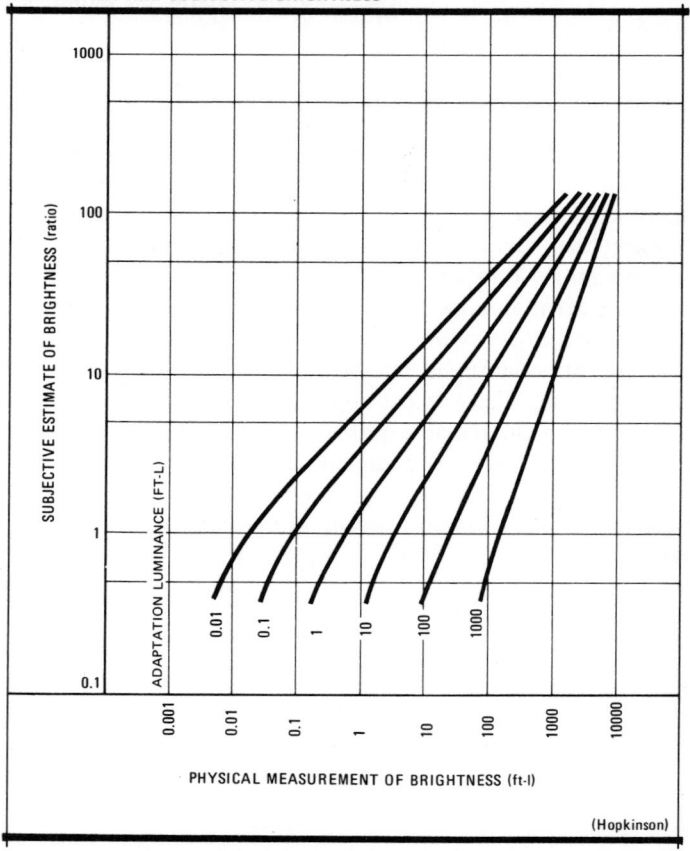

PHYSICAL MEASUREMENT OF BRIGHTNESS (ft-l)

(Hopkinson)

Furthermore, the casual occupant's patterns of attention and circulation may be influenced by his visual attraction to areas of prominent visual contrast. Lacking other incentives, the individual will generally tend to move toward an area of higher brightness.

However, since perception of brightness intensity varies subjectively with the adaptation of the individual eye, it is necessary that transitional measurements be interpreted in this sense. For this purpose, there have been several attempts to develop scales of *apparent* or subjective brightness. And while these efforts must still be regarded as limited in accuracy, they are nevertheless adequate and useful for general estimating purposes.

Figure 1–4.2 shows the results of one series of studies. The diagram relates apparent changes in subjective magnitude (vertical scale) to actual physical changes in intensity (horizontal scale) for several levels of eye adaptation.

Transitional influences related to color

After a period of visual concentration on a small bright form or area, an *after-image* of the same shape and complementary hue will appear for a few seconds. This image is due to temporary overstimulation and fatigue of some of the cones in the retina of the eye. Duration of the fatigue, and therefore the duration and strength of the image, will depend on the brightness of the original form and the length of time it was viewed.

Successive contrast *Successive contrast* is related to this phenomenon. This effect produces a temporary change in the perception of color as the eye moves (after a period of exposure) from one colored surface to another. The perceptual shift is again due to a partial fatigue of some of the cones, increasing the relative sensitivity of the eye to wavelengths *complementary* to the original color.

Successive contrast is also observed after exposure to tinted or saturated colors of light. This action tends to exaggerate spatial color shifts as an individual moves between spaces that are lighted with sources of differing *whiteness*. For example, a visually *warm* space will initially appear much warmer to an individual entering from a visually *cool* space than it will to an individual entering from another *warm* space.

Simultaneous contrast *Simultaneous contrast* is another related visual phenomenon that occurs when a neutral surface is surrounded by color. The neutral panel (particularly white or gray) will appear to be tinted a color *complementary* to the background.

Two or more adjacent colored surfaces viewed simultaneously will affect each other in a similar way; and simultaneous contrast may alter the apparent hue, saturation, and brightness of wall, floor, or ceiling surfaces. However, it should be noted that this effect is significantly reduced when the adjacent areas of color are separated by a narrow neutral area (white, beige, gray, black).

Slow Change or Subliminal Contrast

When environmental conditions change very slowly, the rate-of-change may be *subliminal* (i.e., below the threshold of sensation). When this occurs, the body mechanism is tending to *adapt* or adjust itself to accommodate each successive change, making the subjective effect almost imperceptible. This condition occurs both when the stimulus is increasing in intensity and when it is declining.

As the occupant moves from a bright space to a dim one, therefore, he is immediately conscious of a major change in the sensory environment. However, if a room becomes visually *dim* (or thermally *stuffy*) quite gradually, the occupant may not perceive the change, unless he is exposed to an abrupt renewal of the original condition.

When environmental continuity is required, changes in the intensity of various stimuli should be subliminal.

Perceptible differences in stimulus magnitude

Observations and experiments indicate that minimum perceptible differences in stimulus are not constant intensities. Instead, a *just noticeable difference* is found to be a *ratio* (as described in Weber's Law and the Weber-Fechner Law).

$$\frac{M_2}{M_1} = K$$

where: M_1 = magnitude of the original stimulus
M_2 = magnitude of the new stimulus
K = the applicable minimum ratio (constant)

While this system of ratio relationships is not universally true (particularly in regard to thermal stimuli), it is approximately true in reference to vision, hearing, pressure, and muscular action. In each case, the rule is most accurate in the middle regions of the sensory scale, and becomes more uncertain near the upper and lower limits of perception. So although the principle lacks precision and authority in some respects, it nevertheless serves as a valuable approximation in many phases of environmental design.

The precise minimum ratio will vary with the sense being measured. But for brightness and sound intensity, the approximate minimum meaningful change in intensity is a ratio of two. As a result, general brightness differences between adjacent spaces of less than 2:1 tend to be perceptually insignificant. This means that successive doubling of brightness intensity is the minimum ratio that will produce a sensation of perceptible steps of change.

Similarly, a meaningful change in sonic intensity requires at least a doubling of power level. On the logarithmic decibel scale, this is a minimum change of 3 db (see Table 2–2.2).

Thermal influences are not generally subject to the *constant ratio* rule. However, changes of 3 ET or 6° MRT have been previously identified as perceptible near the comfort zone. Short-term changes of lesser magnitude tend to be subliminal.

PART 2

System Performance and Method

Performance of Materials, Forms, and Individual Systems

Sooner or later, discussions of environmental design must move beyond the subjective environmental criteria, and there occurs the need to select techniques to achieve the intended result. This introduces the problems of equipment and material selection, assembly, and installation; and the need to assimilate and coordinate the physical components into a logical and harmonious building system.

In this regard, the designer is guided by the sense of integrity and logic that must dominate the selection and use of environmental devices.

For example, a beam of light depends on the integrity of the relationship between a light source, a reflector, and a cover plate or facing panel. A change in any one of these will alter the characteristics of the beam of light.

Visual perception of a space (in terms of brightness, form, or color) depends on the integrity of the relationship between direct and reflected lighting elements, specific surface finishes, and the subjective reference level of the observer. A change in any one of these will alter perception.

A subjective sense of acoustical quality depends on the integrity of the relationship between room volume, the nature of surfaces and finishes, the density of room occupancy, and the source-to-background contrast. A change in any one of these will alter sonic perception.

An individual's sensation of thermal comfort depends on the relationship between various heat-producing and heat-consuming elements in a space. The integrity of that relationship depends upon the formulation of a logical order of performance that complements the natural laws of physiology, climatology, and physics, rather than in a generally costly attempt to *overpower* these natural laws in an effort to *save* an undisciplined design.

Designers must be conversant in the use of meaningful devices and materials and in the manipulation of significant relationships. They are fundamental in any attempt to regulate and control the sensory environment.

CHAPTER 2-1

Light Generation and Control

The human environment is subject to a number of natural and man-made sources that emit energy in various regions of the electromagnetic spectrum. This spectrum is shown in Figure 2–1.1.

The segment approximately defined as 3800–7600 angstroms is gener-

FIGURE 2–1.1

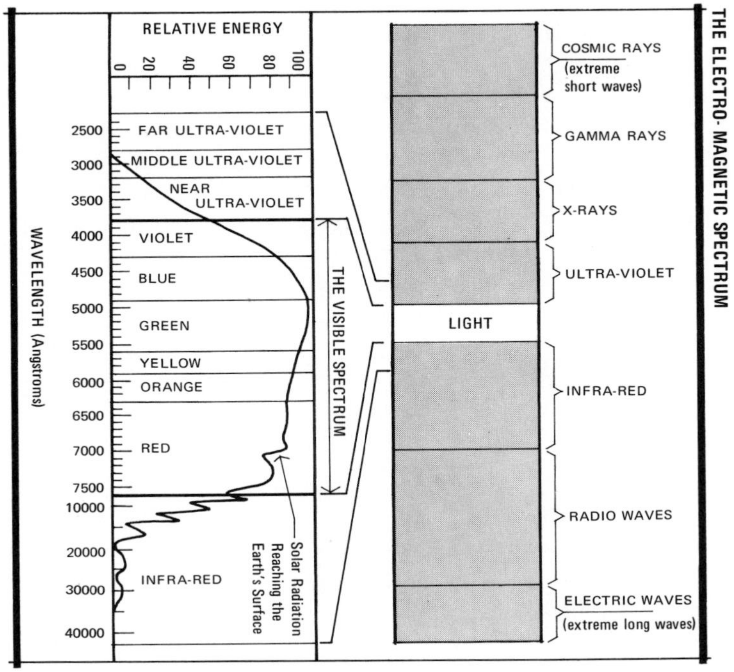

ally referred to as the *visible spectrum* (i.e., the human eye is sensitive to this band of energy). The term light, then, refers to energy emitted in this region.

Since light is a necessary ingredient for normal orientation and activity, spaces for human occupancy must provide for this by utilizing either natural or man-made sources (or both).

CONTROL OF NATURAL LIGHT

The sun is, of course, the dominant lighting influence in nature. All other natural lighting influences and forms perform in response to this source.

In defining the nature of direct solar energy, then, the sun emits a continuous spectrum of energy; and because of the very high source temperature involved, the total spectrum ranges into the longer and shorter wavelengths on both sides of the visible band. As a result, intensity of energy throughout the visible range is fairly uniform.

However, the atmosphere that surrounds the earth modifies solar energy by absorption, reflection, and selective scattering. This action tends to limit the direct penetration of energy in both infrared and ultraviolet regions. It also tends to modify the intensity at various specific wavelengths in the visible region. The water vapor content is particularly important, for this produces transmission changes ranging from clear sky to haze and overcast conditions. These momentary characteristics of the atmosphere, then, tend to produce extremely variable light conditions on earth.

As a further variable, the tilt (declination) of the earth's axis plus the normal daily rotation combine to produce a constantly changing angular relationship between the sun and any specific location on earth. This, in turn, produces changes in solar intensity on both an hourly and a seasonal basis.

Daylight Analysis

In analyzing natural light at a given location, there are three basic components to be considered: (1) direct sunlight, which impinges intermittently on the east, south, or west exposures of buildings in the north temperate region; (2) skylight, which impinges simultaneously and somewhat more consistently on all exposures of a building; and (3) reflected light from the ground and from nearby man-made elements.

Each of these components will vary with time of day, with season, and with the prevalent atmospheric conditions.

Direct sunlight

Direct sunlight is coincident with maximum solar heat. For this reason, it is generally desirable to limit the penetration of this component into the interior.

Table 2-1.1 Typical Daylight Design Data—1

December 22—(40° North Latitude)

	Solar Time	Location of Sun	
		Azimuth[1]	Altitude
Sunrise	7:30	121°0'	0°0'
	8:00	127°0'	5°30'
	Noon	180°0'	26°30'
	4:00	127°0'	5°30'
Sunset	4:30	121°0'	0°0'

[1]*Azimuth* refers to the horizontal angle from north.

	Approximate Average Sky Brightness (foot-lamberts)		
	8 AM	Noon	4 PM
Typical overcast condition	190	930[2]	190
Typical clear day condition N	275	525	275
E	975	875	275
S	750	2175	750
W	275	875	975

[2]A uniformly bright sky of 930 ft-l would produce approx. 930 ft-c on the ground and on all horizontal surfaces.
Vertical illumination from this source is approximately:
    Sky effect: Sky ft-lambert x 0.50.
    Ground egect: Sky ft-lambert x 0.50 x ground reflection.

	Typical Direct Solar Illumination (footcandles)		
	8 AM	Noon	4 PM
Perpendicular[3] to sun rays	2250	6650	2250
Horizontal	100	2850	100

[3]Illumination on a given plane is obtained by multiplying perpendicular illumination times the cosine of the angle of incidence.

In the development of shielding devices, or where limited penetration may be desirable for occasional spatial effect, the action of sunlight can be predicted through graphic projection of solar azimuth and altitude angles. Typical angles are indicated in Tables 2–1.1, 2–1.2, and 2–1.3. Note that these solar angles vary significantly throughout the day and year.

Typical direct solar illumination intensities are also indicated in Tables 2–1.1 thru 2–1.3.

Table 2-1.2 Typical Daylight Design Data—2

March 21, September 21—(40° North Latitude)

	Solar Time	Location of Sun	
		Azimuth[1]	Altitude
Sunrise	6:00	90°0'	0°0'
	8:00	110°30'	22°30'
	Noon	180°0'	50°0'
	4:00	110°30'	22°30'
Sunset	6:00	90°0'	0°0'

[1]*Azimuth* refers to the horizontal angle from north.

		Approximate Average Sky Brightness (foot-lamberts)		
		8 AM	Noon	4 PM
Typical overcast condition		790	1760[2]	790
Typical clear day conditions	N	725	850	725
	E	2450	1475	600
	S	1700	2700	1700
	W	600	1475	2450

[2]A uniformly bright sky of 1760 ft-l would produce approx. 1760 ft-c on the ground and on all horizontal surfaces.
Vertical illumination from this source is approximately:
  Sky effect: Sky ft-lambert x 0.50.
  Ground egect: Sky ft-lambert x 0.50 x ground reflection.

	Typical Direct Solar Illumination (footcandles)		
	8 AM	Noon	4 PM
Perpendicular[3] to sun rays	6050	8250	6050
Horizontal	2050	6000	2050

[3]Illumination on a given plane is obtained by multiplying perpendicular illumination times the cosine of the angle of incidence.

Table 2-1.3 Typical Daylight Design Data—3

June 21—(40° North Latitude)

	Solar Time	Location of Sun	
		Azimuth[1]	Altitude
Sunrise	4:30	59°0'	0°0'
	8:00	89°0'	37°30'
	Noon	180°0'	73°30'
	4:00	89°0'	37°30'
Sunset	7:30	59°0'	0°0'

[1]*Azimuth* refers to the horizontal angle from north.

		Approximate Average Sky Brightness (foot-lamberts)		
		8 AM	Noon	4 PM
Typical overcast condition		1290	3060[2]	1290
Typical clear day conditions	N	1325	950	1325
	E	2850	1400	700
	S	1350	2400	1350
	W	700	1400	2850

[2]A uniformly bright sky of 3060 ft-l would produce approx. 3060 ft-c on the ground and on all horizontal surfaces.
Vertical illumination from this source is approximately:
   Sky effect: Sky ft-lambert x 0.50.
   Ground egect: Sky ft-lambert x 0.50 x ground reflection.

	Typical Direct Solar Illumination (footcandles)		
	8 AM	Noon	4 PM
Perpendicular[3] to sun rays	7600	8850	7600
Horizontal	4550	8100	4550

[3]Illumination on a given plane is obtained by multiplying perpendicular illumination times the cosine of the angle of incidence.

Skylight

Although still variable in intensity and color, skylight tends to be a more consistent source of natural daylight.

But skylight must be evaluated in two different forms: (1) the overcast sky condition, and (2) the clear sky condition. Both of these are similar in that they represent a large-area diffuse light source (as contrasted with the sun, which functions environmentally as a high intensity, small-area source).

The two forms of skylight differ in intensity, in color quality, and in directional uniformity (see Tables 2–1.1 thru 2–1.3, center). The method for estimating the relative intensity of skylight on vertical and horizontal building surfaces is shown in note 2 in Table 2–1.1.

Reflected light from the ground

The method for estimating the relative intensity of *ground light* on a vertical building surface is also shown in Table 2–1.1 (note 2).

Since ground surfaces are subject to control by the architect, the selection of adjacent landscaping materials becomes a moderate control device to influence the intensity of daylight that is incident on the building shell (see Table 2–1.4). Furthermore, since much of the light that is reflected from the ground is directed toward the interior ceiling surface, the interior distribution and intensity of daylight in low buildings can be manipulated through the interaction of these two reflecting surfaces (ground and ceiling).

Table 2-1.4 Typical Daylight Design Data (Ground Reflectance)

	Typical Reflectance
Grass	6%
Vegetation (mean value)	25%
Earth	7%
Snow: new	75%
old	60%
Concrete	55%
White marble	45%
Brick: buff	45%
red	30%
Gravel	13%
Asphalt	7%
Painted surface: new white	75%
old white	55%

For low buildings on a sunny day, light reflected from the ground will typically approach 50% of the total daylight incident on a window area that is shaded from direct sunlight (see Figure 2–1.2, curve 2). It may

FIGURE 2–1.2

TYPICAL DAYLIGHT PERFORMANCE (ATMOSPHERIC VARIATION)

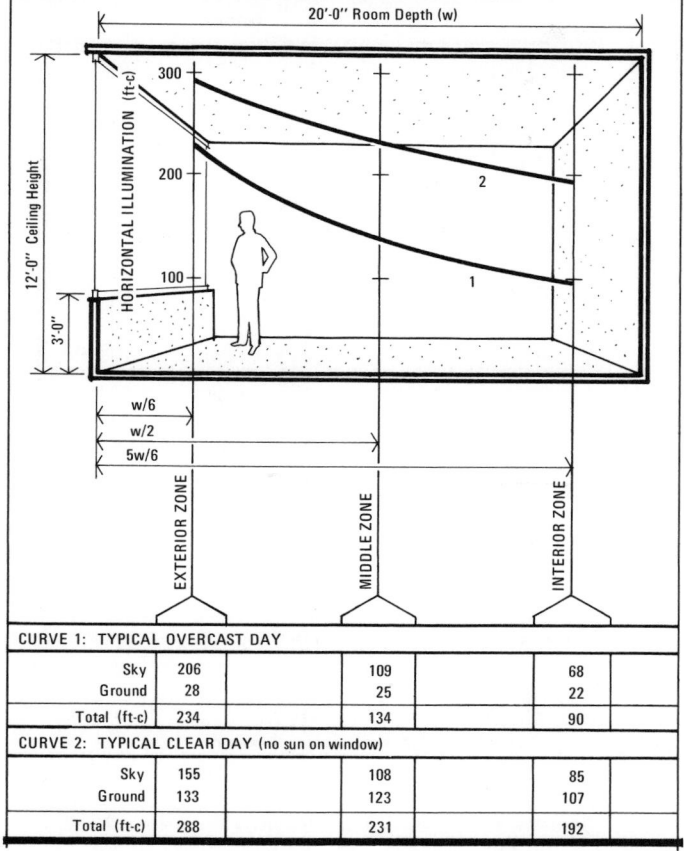

CURVE 1: TYPICAL OVERCAST DAY							
Sky	206		109		68	
Ground	28		25		22	
Total (ft-c)	234		134		90	

CURVE 2: TYPICAL CLEAR DAY (no sun on window)							
Sky	155		108		85	
Ground	133		123		107	
Total (ft-c)	288		231		192	

exceed this proportion when the building is surrounded by high reflectance surfaces, such as sand, light concrete, or snow cover.

For overcast conditions, the light reflected from the ground will be less significant, typically representing approximately 10–25% of the daylight incident on the window.

Man-Made Enclosures

Throughout history, building design has involved a paradox. There is the need to provide structures that are adequately sealed to prevent the intrusion of adverse influences such as cold winds, rain, etc.; and yet these

FIGURE 2–1.3

TYPICAL DAYLIGHT PERFORMANCE (GROUND VARIATION)

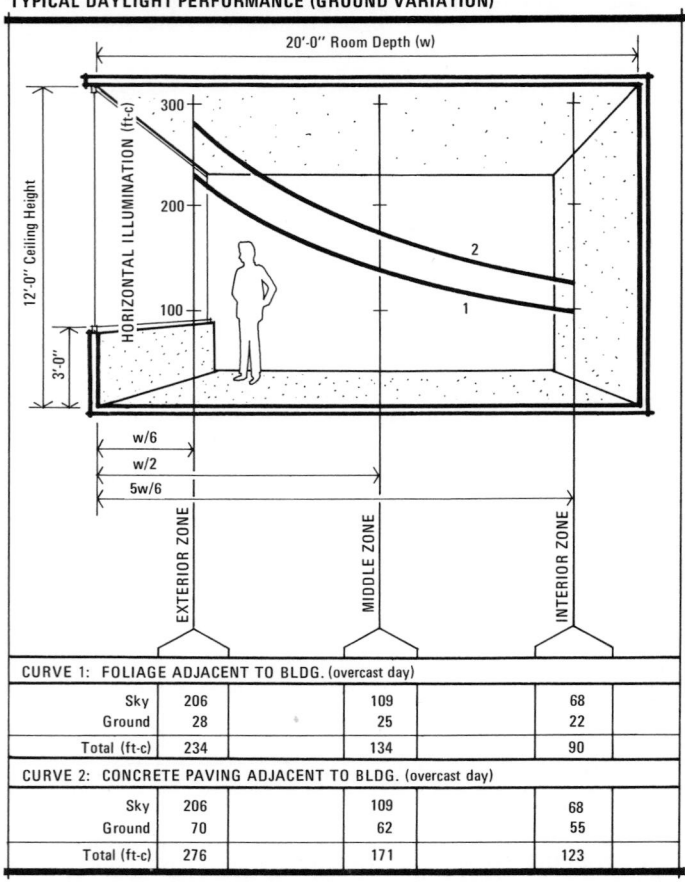

CURVE 1: FOLIAGE ADJACENT TO BLDG. (overcast day)						
Sky	206		109		68	
Ground	28		25		22	
Total (ft-c)	234		134		90	
CURVE 2: CONCRETE PAVING ADJACENT TO BLDG. (overcast day)						
Sky	206		109		68	
Ground	70		62		55	
Total (ft-c)	276		171		123	

same structures must be adequately open to permit the penetration of light and ventilation air. Each society and culture has produced its own solution to this problem, reflecting the specific demands of the region, the technology of the time, and the sensitivity of the people.

Early western civilizations developed in the arid and semiarid regions near the 70° isotherm. In these regions, the natural lighting condition is both intense and seasonally consistent. Minimal openings in Egyptian and Greek structures, then, were often sufficient to permit adequate light to penetrate into the interior for purposes of spatial definition and orientation. Similarly, exterior sculptural detail and building form were relatively subtle because the intense, directional light is sufficient to produce sharp shadows for contrast.

In medieval Europe, Gothic and Baroque architecture evolved more open structures to facilitate penetration of the more gentle and variable daylight that is associated with temperate climates. At the same time, stonework tracery was developed to provide more richly molded forms because natural shadows tend to be more subtle.

Building in the temperate zones remain subject to wide variations in the character of daylight—particularly in the northern and middle areas of the United States, where *diffuse* daylight conditions tend to predominate during a majority of the days in a year.

The following comments relate to buildings located in north temperate areas. They summarize some of the more basic physical relationships involved in utilization of daylight in this region.

Orientation of building faces

Assuming suitable sun control, the south face of the building generally affords the maximum quantities of light. This is particularly true during the winter months (see Tables 2–1.1 thru 2–1.3).

Because of the low angles of the sun during the morning and afternoon hours, the east and west exposures present the most difficult problems related to direct sunlight penetration and glare control. (These same exposure conditions tend to complicate the thermal environment. Screening techniques are discussed in Chapter 2–3.)

Except for very short early morning and late afternoon intervals during the peak of the summer solstice, the sun does not impinge on the north face of buildings in the northern temperate regions. For this reason, skylight and reflected light from the ground are the dominant sources of north daylight; and this light remains relatively consistent throughout the day. North exposure is useful, then, when consistency of daylight color and shadow effects is required.

Development of building form

Since light intensity diminishes with distance from the source, the penetration capability of daylight is limited. For this reason, a decision to utilize daylight will tend to limit the building form to those configurations which permit the introduction of light openings near the significant interior task centers.

This may lead to relatively narrow building configurations that permit light to enter from either or both sides; it may lead to low buildings that permit light to enter through roof openings. It may also lead to the use of courtyards and *light wells* to facilitate the penetration of daylight.

Window elements   Window elements afford the occupants a view of the outdoor environment. This concept of a transparent screen has obvious significance for assisting the development of a total sense of orientation and environmental relationship (*outlook*).

As a source of light (to be distinguished from this function of a transparent screen), long low windows will tend to provide a somewhat elliptical distribution of light in the interior. There will be a moderate intensity of light distributed to each side of the opening.

FIGURE 2–1.4

TYPICAL DAYLIGHT PERFORMANCE (CEILING HEIGHT)

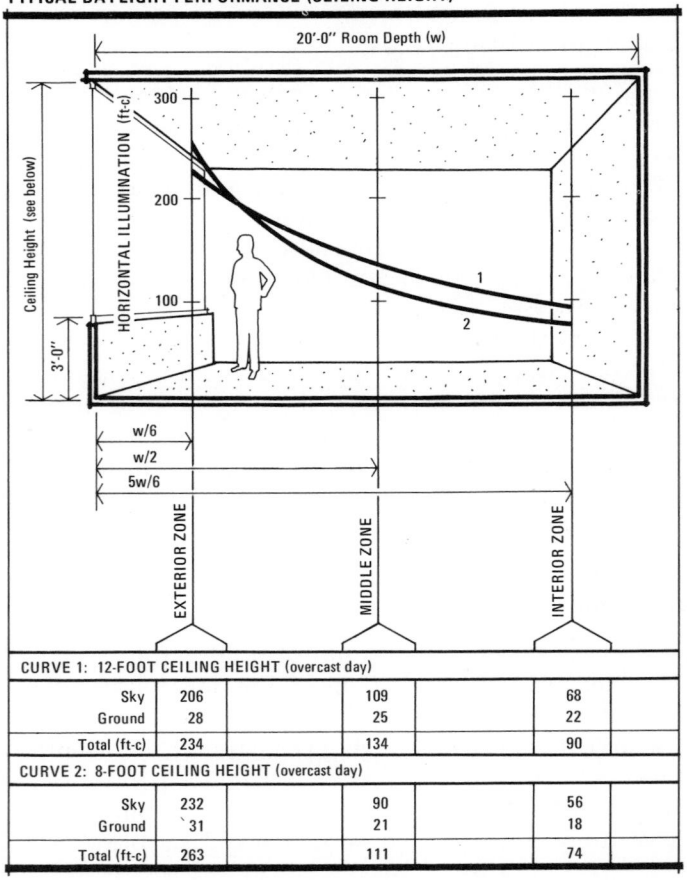

CURVE 1: 12-FOOT CEILING HEIGHT (overcast day)						
Sky	206		109		68	
Ground	28		25		22	
Total (ft-c)	234		134		90	
CURVE 2: 8-FOOT CEILING HEIGHT (overcast day)						
Sky	232		90		56	
Ground	31		21		18	
Total (ft-c)	263		111		74	

A high, narrow window will, on the other hand, provide more depth of penetration, but with relatively little distribution of light to each side.

Window Height and Room Depth.   When a sense of general brightness consistency is desired from window elements, the effective depth (width) of rooms in more northern temperate regions should be limited to a maximum of 2–2½ times the height from the floor to the window head. This assumes the use of continuous or near-continuous window elements.

When this ratio is exceeded and overcast days are common, the general brightness of the outermost zone in the room usually exceeds the brightness of the innermost zone by a factor greater than 4 or 5:1. These more severe gradients tend to be excessive when an impression of general consistency is desired.

FIGURE 2–1.5

TYPICAL DAYLIGHT PERFORMANCE (BILATERAL SECTIONS)

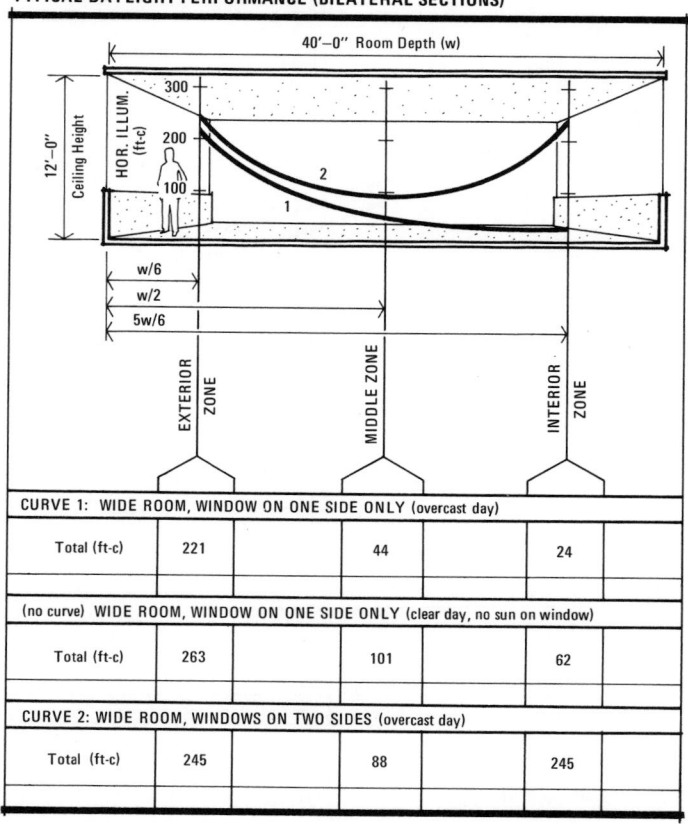

CURVE 1: WIDE ROOM, WINDOW ON ONE SIDE ONLY (overcast day)		
Total (ft-c) 221	44	24

(no curve) WIDE ROOM, WINDOW ON ONE SIDE ONLY (clear day, no sun on window)		
Total (ft-c) 263	101	62

CURVE 2: WIDE ROOM, WINDOWS ON TWO SIDES (overcast day)		
Total (ft-c) 245	88	245

However, in regions where clear days are more prevalent and reliable (such as the southwestern United States), the gradients tend to be less severe. Note the analysis of the wide room (clear day) in Figure 2–1.4. In these clear atmospheric conditions, room width to window height ratios may be increased to 3–3½ times without exceeding the previously-noted brightness gradients.

Bilateral Sections. When the previous room width to window height ratios cannot be maintained, one option which will permit the development of wider room sections (for a given ceiling height) is the addition of window openings in the opposite wall (see Figure 2–1.5). This may be a full

FIGURE 2–1.6

TYPICAL DAYLIGHT PERFORMANCE (SURFACE REFLECTANCE)

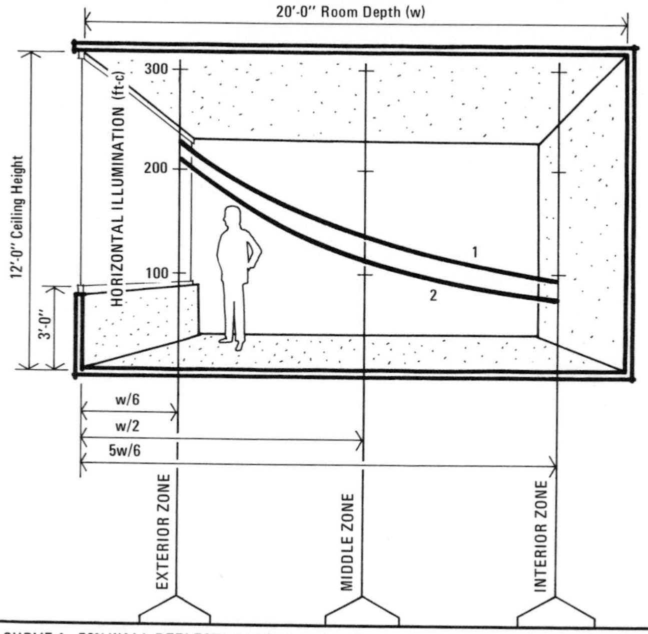

CURVE 1: 70% WALL REFLECTANCE (80% ceiling, 30% floor)					
Sky	206		109		68
Ground	28		25		22
Total (ft-c)	234		134		90
CURVE 2: 30% WALL REFLECTANCE (80% ceiling, 30% floor)					
Sky	187		93		47
Ground	24		19		15
Total (ft-c)	211		112		62

or partial window section, and may also contribute to the natural ventilation of the space by providing cross-ventilation.

Since at least one of the window elements will probably face the sun during significant portions of the day, some type of sun control device will likely be required.

Interior Surface Reflectances. Generally, the most significant single interior surface for assisting the interior reflection of daylight is the ceiling. When penetration and improved spatial uniformity is desired, this surface should be a diffuse, high reflectance finish (approximately 60–80%). Higher reflectance floor finishes (approximately 30%) are also significant in the sense that they also assist the interreflection of light.

Where ceiling and floor reflectances are high, the further influence of wall finishes is indicated in Figure 2–1.6.

Sky Glare. Openings that are intended to facilitate the penetration of daylight are also likely to become sources of sky glare and sun glare. This may necessitate the use of a brightness control device, such as external baffles, venetian blinds, glare-reducing glass, draperies, or tree landscaping. The resulting screen action will, of course, also tend to restrict the penetration and intensity of light in the interior.

(Also see Chapter 2–3 for a more specific discussion of solar screens.)

Clerestories and roof monitors Clerestories and roof monitors (Figure 2–1.7) are one category of forms that may be useful for overcoming the previously discussed limitations in the room width to window height ratio. Specifically, these devices provide a means for emitting light into the more remote interior portions of spaces that are located immediately below the roof.

Combination Systems. When window walls are supplemented by clerestory windows, both openings should be oriented in the same direction (see Figure 2–1.8).

The setback distance from the window (S_1) should be approximately 1–1½ times the window height. When continuous or near-continuous windows are involved, this setback can be increased to approximately 2 times window height.

When clerestory sill heights are relatively low, the height of the clerestory opening should be approximately one-half of the side wall window height (H_c = minimum ½H_w). When the clerestory sill is high (i.e., approx. 2½–3 times the side wall window height), the height of the clerestory opening should be approximately equal to the window height.

The distance from the clerestory window to the *dark* far wall (S_2)

FIGURE 2–1.7

BASIC CLERESTORY AND MONITOR SECTIONS

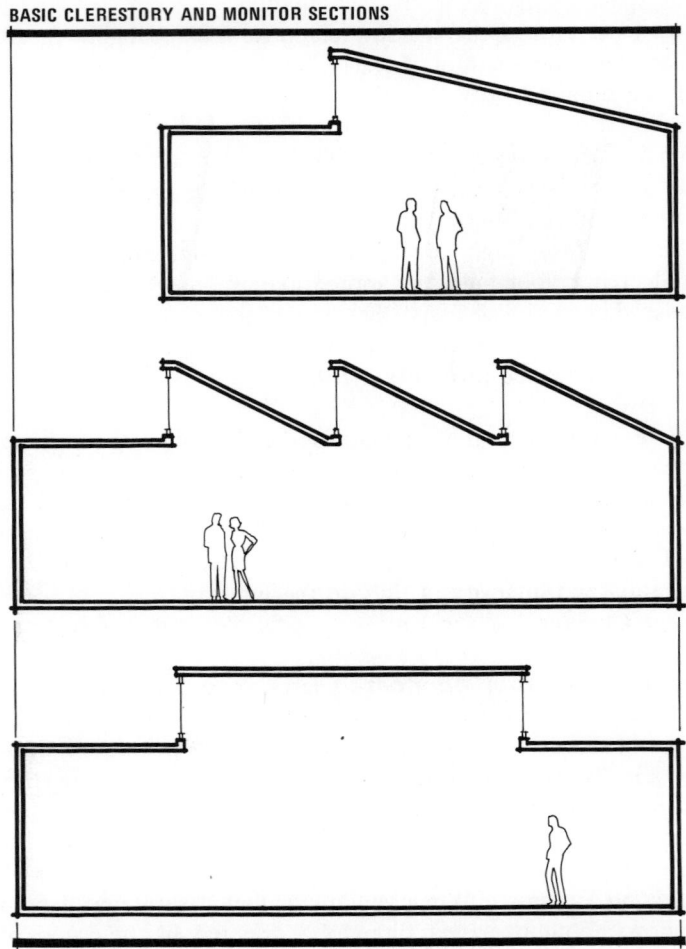

should not exceed two times the height to the top of the clerestory (H_t). For small clerestory openings (H_c), this distance should be reduced to 1½ times.

When greater than manageable S_2 distances are involved, consider the monitor roof.

Roof openings Roof openings can also provide daylighting for deep or isolated interior spaces. The principal disadvantage is the fact that

FIGURE 2–1.8

COMBINATION SYSTEM PROPORTIONS

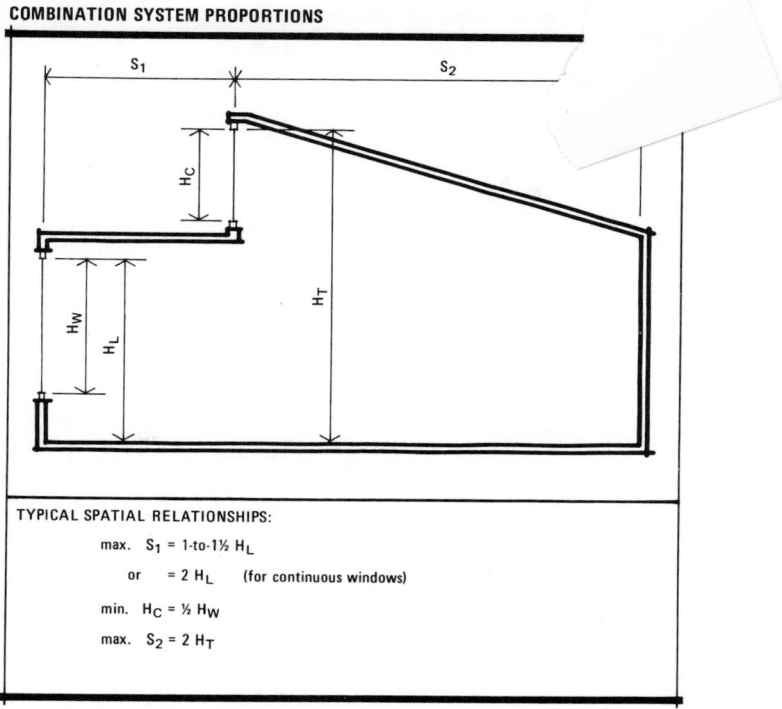

TYPICAL SPATIAL RELATIONSHIPS:

max. S_1 = 1-to-1½ H_L

or   = 2 H_L (for continuous windows)

min. H_C = ½ H_W

max. S_2 = 2 H_T

major solar heat loads generally impinge on the horizontal roof surface, and these loads will penetrate directly into the interior space through the openings.

This problem of solar penetration is a particularly significant consideration for roof openings because direct solar loads are incident on the roof continuously over most of the day, rather than the periodic peaks that occur for individual wall orientations. For this reason, external canopies, light wells, and other shielding devices are sometimes used to reduce the direct solar penetration through roof openings (see Figure 2–1.9).

Regarding the placement of openings, when the system is intended to produce uniform brightness on an interior horizontal surface, moderate-sized openings should be placed so that the center to center spacing does not exceed two times the floor to ceiling height (or, for work areas, two times the height of the ceiling above the work plane). Small-area roof openings provide a much more narrow distribution. For this reason, the maximum center to center spacing of smaller openings should not exceed the floor to ceiling distance when horizontal uniformity is desired.

FIGURE 2–1.9

CANOPY SHIELDS FOR SKYLIGHT ROOF OPENINGS

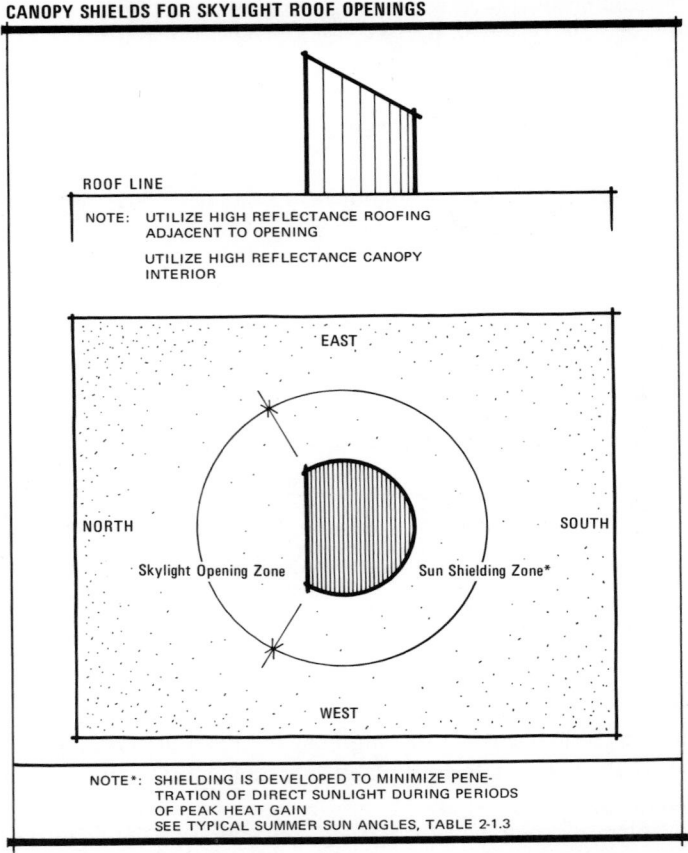

Opening Efficiency. The effectiveness of shielded openings (light wells) will vary (1) with the size of the well opening; (2) with the depth of the well; and (3) with the reflectance of the well walls. These factors can be related with the following formula:

$$I_w = \frac{H_w \times (W_w \times L_w)}{2(W_w \times L_w)}$$

where: I_w = well index (for use in Figure 2–1.10)
W_w = width of well opening
L_w = length of well opening
H_w = depth of well opening

For general estimating purposes, Figure 2–1.10 further relates these factors and indicates the relative transmission efficiency of various well configurations. The transmission of the glazing (including the anticipated effect of dirt accumulation) will further reduce this efficiency.

FIGURE 2–1.10

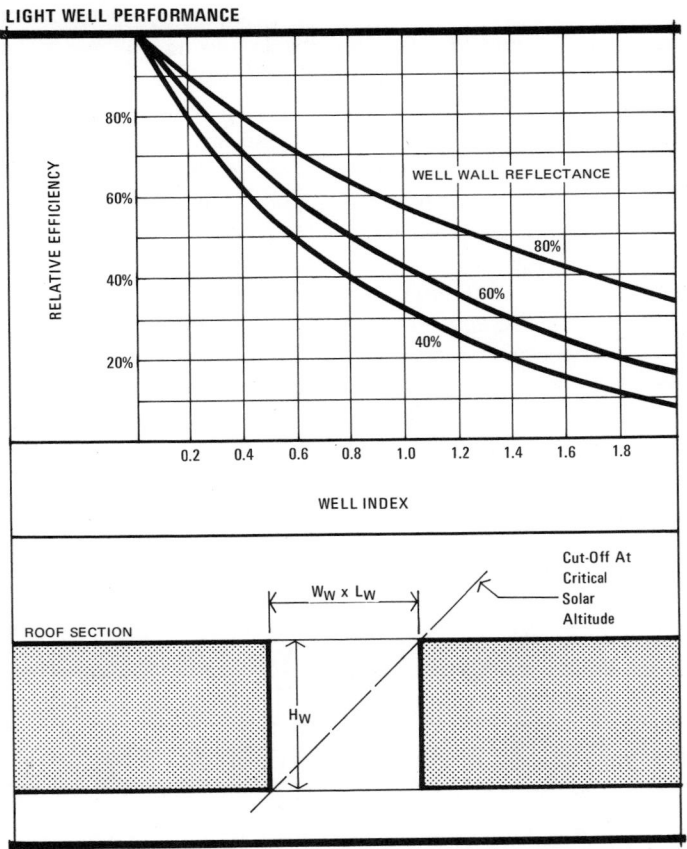

CONTROL OF ELECTRIC LIGHT

The open fire, the torch, the candle, the oil and gas lanterns—each of these satisfied, at least in part, man's need for a source of light that he could readily manipulate and control. The electric lamp is a continuation of this line of development.

In the evaluation of these man-made sources, it should be noted that all heated solids will emit a *continuous spectrum* of energy. This implies an emission of all wavelengths over a certain portion of the electromagnetic spectrum.

However, the radiant energy emitted by the solid does not involve the same intensity at all wavelengths. Furthermore, as the absolute temperature of the solid increases, the energy peak will shift toward the shorter-wave end of the spectrum.

As an example, a solid that is subjected to the normal human environment of 70–80°F (530–540°K) will generally emit a radiant energy peak at about 100,000 angstroms. Since this energy is not visible to the eye, we call this solid *black*. On the other hand, an incandescent lamp tungsten filament operating at a solid temperature of 2900°K will peak at about 10,000 angstroms in the near infrared region, with some of the continuous spectrum now being emitted in the visible region below 7600 angstroms (see Figure 2–1.11, top).

In the normal range of environmental room temperatures, therefore, objects and surfaces will emit energy in the far infrared region. But as the temperature of the object increases, portions of the visible spectrum are also ultimately emitted. As this action takes place, the human observer first perceives a red glow; and then, as the object temperature continues to increase, the observer perceives a *white* glow. This relationship of color and temperature is described by the action of a theoretical *blackbody radiator,* as summarized in Table 2–1.5. (The action of tungsten is near the theoretical action shown.)

Table 2-1.5 Blackbody Radiation

Object Temperature	Visible Color Emitted by the Object
Room temperature	Black (No visible emission)
800°K	Red
3000°K	Yellow
5000°K	White
8000°K	Pale blue
60,000°K	Deep blue

A second major method of light production involves energy that is emitted by a gas, when that gas is subjected to an electric discharge. In this case, radiation is emitted as a *discontinuous line spectrum.* This implies an emission that is made up of clearly definable concentrations at certain wavelengths, separated by regions in which there is no radiation.

In the initial emission of energy, a fluorescent lamp produces a line spectrum of energy. However, with the extremely low gas pressures found in a fluorescent lamp (approximately 1/100,000 of an atmosphere), this initial energy is emitted almost completely in the ultraviolet portion of the spectrum. In order to become a relevant lighting device, then, the fluorescent lamp must rely on the interaction of the emitted ultraviolet with the phosphors that coat the inside of the bulb. These phosphors convert the UV energy to longer wavelengths, and thus cause the device to emit a continuous spectrum of light (see Figure 2–1.11, bottom).

Mercury vapor and other high pressure discharge lamps also produce a

FIGURE 2–1.11

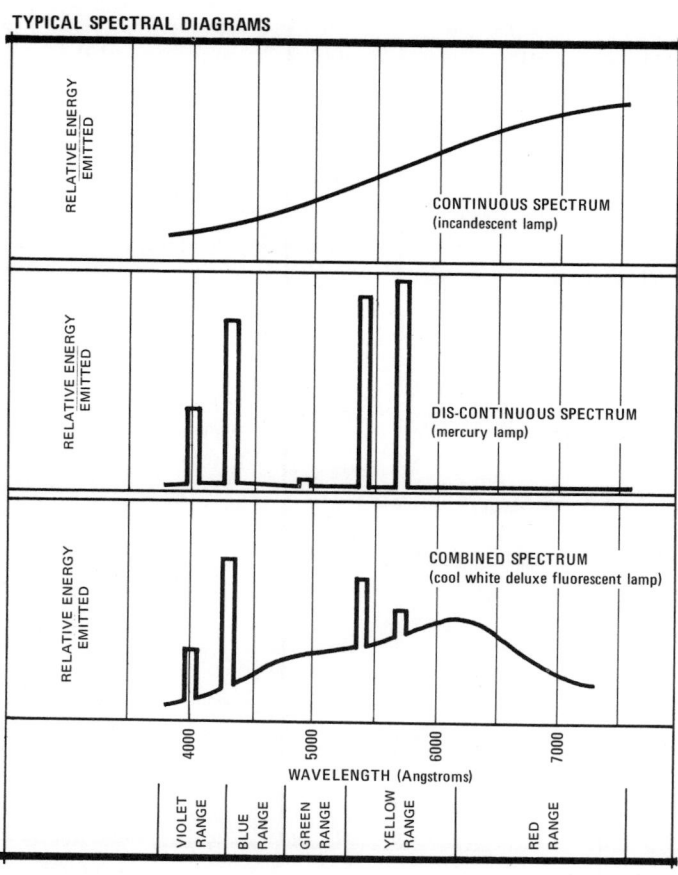

TYPICAL SPECTRAL DIAGRAMS

RELATIVE ENERGY EMITTED

CONTINUOUS SPECTRUM
(incandescent lamp)

RELATIVE ENERGY EMITTED

DIS-CONTINUOUS SPECTRUM
(mercury lamp)

RELATIVE ENERGY EMITTED

COMBINED SPECTRUM
(cool white deluxe fluorescent lamp)

4000   5000   6000   7000

WAVELENGTH (Angstroms)

VIOLET RANGE   BLUE RANGE   GREEN RANGE   YELLOW RANGE   RED RANGE

discontinuous line spectrum of energy. In this sense, they operate in a manner that is similar to the action of a fluorescent lamp. However, in mercury-type sources, higher operating vapor pressures shift a larger proportion of the initial emission from ultraviolet to longer visible wavelengths (see Figure 2–1.11, center). If this pressure action is extended, at extremely high gas pressures there is a further tendency to broaden the individual line spectra into bands, and thus approach a continuous spectrum.

FIGURE 2–1.12

COLOR 'WHITENESS' (NATURAL - ELECTRIC COMPARISON)

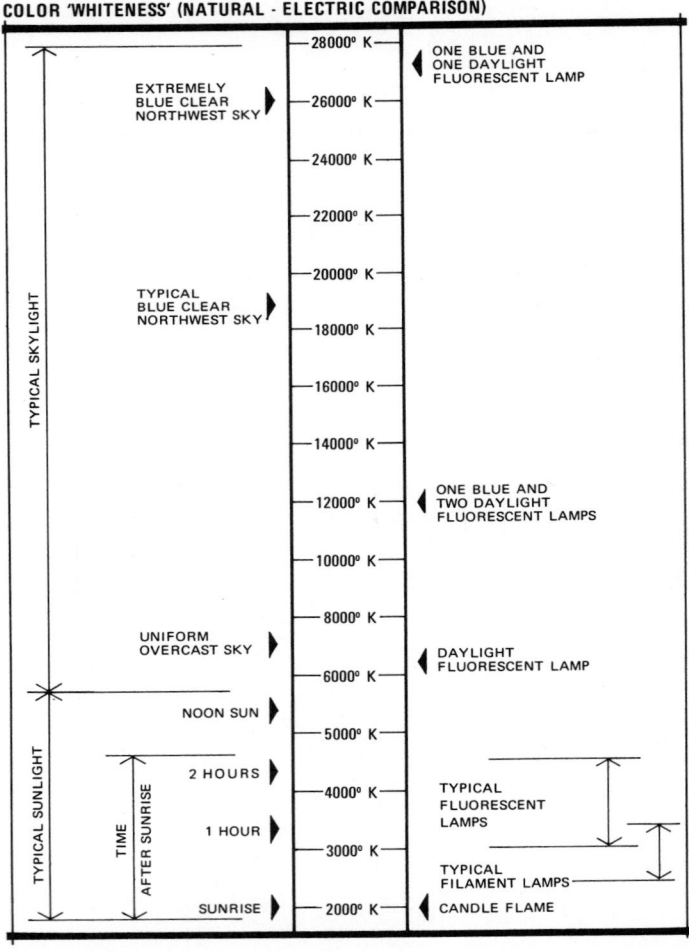

Color compatibility of electric and natural light

Daylight is quite variable in *whiteness,* as is indicated in Figure 2–1.12. However, during most of the day, the mixture of sunlight-plus-skylight will produce a whiteness range that falls between 4000°K and 5500°K (equivalent blackbody temperature). Heavy overcast conditions will produce somewhat higher color temperatures.

This range is generally defined as *cool.* As a result, for interior spaces that involve significant glass areas, the electric system should generally utilize *cool* sources that are compatible with the spectral characteristics of the most prevalent daylight condition.

Where association with daylight is not a factor, warmer light sources are alternatives for consideration.

Material Action

Energy in the electromagnetic spectrum travels at a speed of 186,000 miles/sec and moves in a straight line. If this energy is to bend or otherwise change its basic character, this change must be induced through interaction with various materials and forms.

The conscious manipulation of energy in this sense involves the art and science of *optics.* This, in turn, involves the physical manipulation of light by one or both of two methods: (1) reflection, and (2) transmission.

Reflection

Reflecting materials can be classified as (1) *specular,* like a mirror or polished aluminum, or (2) *diffusing,* like blotting paper or mat paint.

Independent of this, a reflecting material can be classified as *high reflectance* or *low reflectance.* This characteristic refers to the per cent of incident energy that is reflected in any and all manners from the surface. It should be emphasized, however, that *reflectance* is quite independent of the *specularity* or *diffusion* associated with the surface.

Specular reflection   The primary characteristic of specular reflection is the fact that the angle of reflected energy will equal the angle of incidence. Intensity of reflection will be diminished to the degree that the reflectance is less than 100%.

With this knowledge, the designer can predict the action of reflected light. Similarly, where specular planes such as polished walls and table

Table 2-1.6 Electric Light Source Characteristics

	Filament	Fluorescent	High Pressure Discharge
Light form	High intensity *point* source of light. Capable of long-range projection as a directional *cone*	Low intensity *linear* source of light. Capable of short-range projection as a directional *wash*	High intensity *point* source of light. Capable of long-range projection as a directional *cone*
Luminaire size	Compact	Bulky	Moderately compact
Auxiliaries	Socket (1)	Sockets (2) Ballast (for starting and current control)	Socket (1) Ballast (for starting and current control)
Efficiency of light production	10–20 lumens/watt	60–75 lumens/watt	Mercury: 60 lumens/watt Lucolox: 100 lumens/watt
Effective life	1000 hours Varies with voltage (Reduced voltage extends lamp life) PAR, R, QUARTZ types: typically 2000 hours PAR-Q types: typically 4000 hours	10,000–15,000 hours Varies with number of starts (Above estimate based on 3 burning hours/start; greater frequency decreases lamp life)	6000–24,000 hours Varies with number of starts (Above estimate based on 6–9 burning hours/start; greater frequency decreases lamp life)
Maintenance	Average 80–85% of initial output	Average 85% of initial output	Average 80–90% of initial output

	(Decrease is due to condensation of evaporated tungsten on bulb, causing blackening of bulb)	(Decrease is due to phosphor decay; and condensation of evaporated electron-emissive material on bulb, causing blackening of bulb)	(Decrease is due to condensation of evaporated electron-emissive material on bulb, causing blackening of bulb)
Starting and warm-up	Near instantaneous	Fair-to-good	Slow
Dimmability	Good	Good for rapid start types	Cannot dim electrically
Color *whiteness*	Warm	Warm or cool	Typically cool
Color temp., °K	2500–3400°K (see Figure 2-1.12)	2900–7000°K (see Figure 2-1.12)	Mercury, clear: Cool, greenish-white
Associated spatial effect	Yellow-White	Variable, depending on choice of *warm* or *cool* source	Mercury, w/phosphor: Cool, yellow-green to bluish-white. Lucolox: Warm, yellow-white
Color rendition	Continuous spectrum (see Figure 2-1.11) Deficient in blue	Combination spectrum (see Figure 2-1.11)	Discontinuous spectrum (see Figure 2-1.11) Mercury: Deficient in red Lucolox: Deficient in blue
Saturated color	Excellent red, yellow Poor blue, green	Excellent green, blue, yellow Poor red	Mercury: Excellent blue, green, negligible red Lucolox: Excellent yellow, poor blue

tops are involved, the designer can identify the limits of the reflected field of view for each occupant position (see Figure 1–1.21 and 1–1.22).

When specular materials are curved in a convex manner about a compact light source, a focusing action begins to occur for the reflected light. This action can be developed to produce either (1) a narrow, concentrated beam, or (2) a broad, diverging beam from a relatively small opening (see Figure 2–1.13). Specular reflectors are therefore typically used for luminaire reflector devices.

FIGURE 2–1.13

SPECULAR REFLECTION

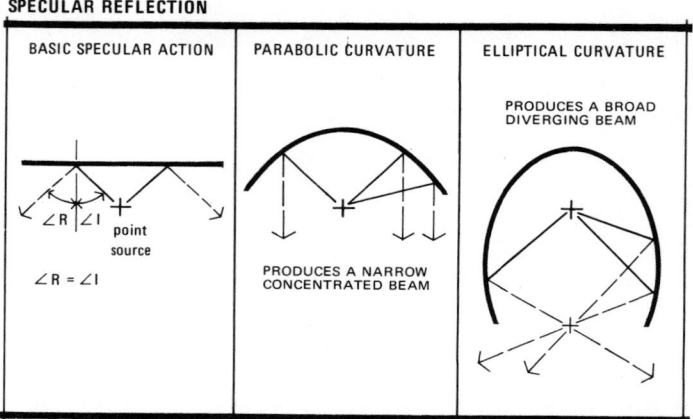

Diffuse reflection The primary characteristic of diffuse reflection is the fact that the angle of reflected energy has little relationship to the angle of incidence. The directional integrity of the beam is destroyed and the energy is scattered (see Figure 2–1.14). Intensity of reflection will again be diminished to the degree that the reflectance is less than 100%.

With this knowledge, the designer knows that a surface of this type will reflect light broadly within a space, and that the subjective view of the observer will be that of a somewhat uniform, integrated surface brightness. For this reason, diffuse materials are typically used for most spatial surfaces (such as walls, etc.).

When diffusing materials are curved, the curved form serves a function of limiting the distribution of brightness slightly; but there is little or no focusing effect. Regardless of form, then, diffuse reflectors in a luminaire will produce a broad or diverging beam, generally from a relatively large opening.

FIGURE 2–1.14

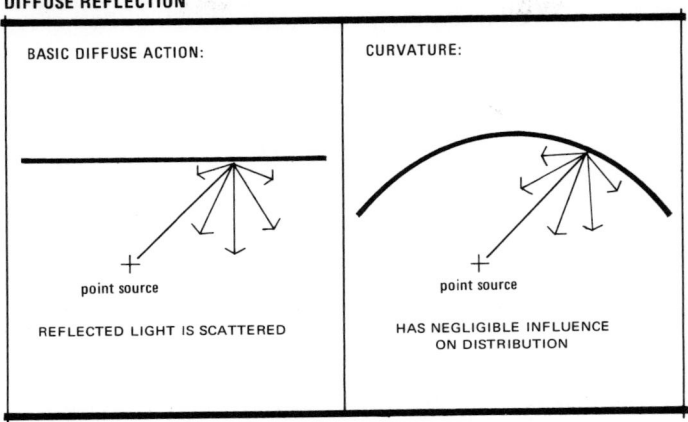

DIFFUSE REFLECTION

BASIC DIFFUSE ACTION:

CURVATURE:

point source

point source

REFLECTED LIGHT IS SCATTERED

HAS NEGLIGIBLE INFLUENCE
ON DISTRIBUTION

Transmission

Transmitting materials can be classified as (1) *clear transparent;* (2) *prismatic,* like clear ribbed glass or plastic; or (3) *diffusing,* like white glass or plastic.

Independent of this, a transmitting material can be classified as *high transmission* or *low transmission.* This refers to the per cent of incident energy that is transmitted in any and all manners through the material.

Prismatic action   Although rays of energy may pass directly through a transparent medium like air, clear water, clear glass, and clear plastic without changing their essential character, the speed of the energy is different in each of these transparent materials. Because of this difference in speed, there is a slight modification in the direction of the energy ray. This directional change occurs at the surface (or surfaces) where the change in speed takes place.

This phenomenon can be observed by placing a straight stick into a clear pool of water. The stick appears to bend at the juncture of the water and air.

If a piece of clear glass has parallel faces, therefore, light will bend at the upper surface, while a compensating bend will occur at the lower surface (see Figure 2–1.15, left). As a result, there is a slight displacement of the beam, but no permanent change of direction. This type of panel is useful when enclosure is required but no change in beam direction is desired (as is required for spotlight cover plates or for window glass).

Table 2-1.7 Light Reflectance (Interior)

	Reflectance
Typical Specular Materials	
Luminaire reflector materials:	
Silver	90–92%
Chromium	63–66%
Aluminum: Polished	60–70%
Alzak polished	75–85%
Stainless steel	50–60%
Building materials:	
Clear glass or plastic	8–10%
Stainless steel	50–60%
Typical Diffusing Materials	
Luminaire reflector materials:	
White paint	70–90%
White porcelain enamel	60–83%
Masonry and structural materials:	
White plaster	90–92%
White terra-cotta	65–80%
White porcelain enamel	60–83%
Limestone	35–60%
Sandstone	20–40%
Marble	30–70%
Gray cement	20–30%
Granite	20–25%
Brick: Red	10–20%
Light buff	40–45%
Dark buff	35–40%
Wood:	
Light birch	35–50%
Light oak	25–35%
Dark oak	10–15%
Mahogany	6–12%
Walnut	5–10%
Paint:	
New white paint	75–90%
Old white paint	50–70%

However, if a piece of clear glass involves nonparallel faces, the light emerges at the lower face with a permanent change in direction (see Figure 2–1.15, center). This prismatic action is a basic principle in lens design.

FIGURE 2–1.15

TRANSMISSION

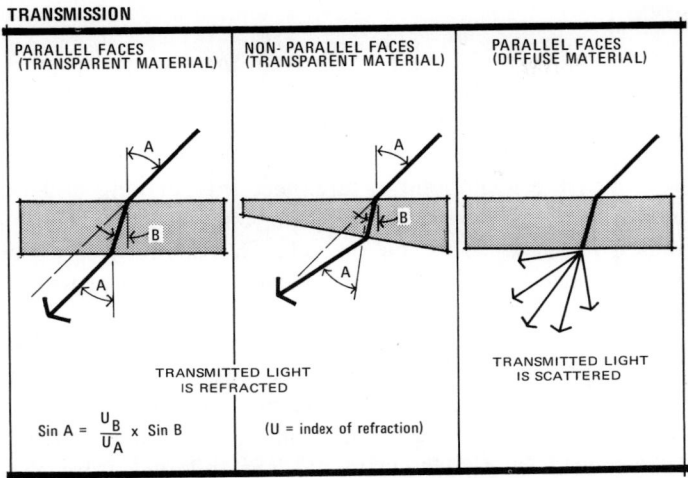

Sin A = $\frac{U_B}{U_A}$ x Sin B   (U = index of refraction)

Diffuse transmission   Diffuse transmission, like diffuse reflection, produces a basic change in an incident ray of light. Regardless of the directional character of the incident ray, the result is a broad scattering of light (see Figure 2–1.15, right).

Also similar, the subjective view of the observer will be that of a uniform, integrated surface brightness. For this reason, diffuse materials are typically used for most moderate and large-area transilluminated surfaces.

Table 2-1.8 Light Transmission (Interior)

	Transmission	*Reflectance*
Clear glass or plastic	80–94%	6–10%
Transparent colors: Red	8–17%	
Amber	30–50%	
Green	10–17%	
Blue	3– 5%	
Configurated or prismatic glass	55–90%	10–25%
Acid-etched glass:		
Toward source	80–90%	5–10%
Away from source	65–80%	10–20%
Opal white glass	12–40%	40–80%
White plastic	30–65%	30–70%
Marble	5–40%	30–70%

Shielding

Louvers and baffles screen direct light from the eye (see Figure 2.1.16). In this sense they help control direct glare and become a basic tool for manipulating luminaire brightness. Furthermore, since it is generally desirable to limit view of the mechanics of the system, shielding may also be justified from the standpoint of architectural detailing and aesthetics.

In general, *baffles* provide shielding in one direction (i.e., along a single viewing axis). *Louvers,* on the other hand, are a series of baffles or shielding elements arranged in a geometric pattern to provide shielding from many directions.

In both cases, shielding is effective within a specified *shielding zone.* This refers to the maximum angle that the eye can be raised above the horizontal without seeing through the shielding system.

Within the shielded zone itself, the brightness of baffle and louver surfaces is determined by the intensity of light reflected from the surface toward the eye. So control of this intensity (to produce high or low system brightness) can be manipulated by varying the reflectance of the louver surface. (Also see Table 1–1.6).

FIGURE 2–1.16

SHIELDING WITH BAFFLES AND LOUVERS

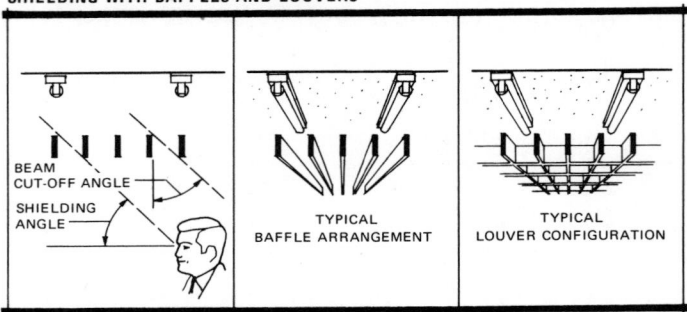

Beam performance

There is a sense of logic in the coordinated use of materials and energy sources. In this sense, a specific beam of light depends on the integrity of the relationship between: (1) reflector forms and finishes; (2) the light source; and (3) the cover plate or facing material. A change in any one of these will alter the characteristics of the beam of light.

The material-source relationships most commonly used for interior environmental systems are summarized in Figures 2–1.17A, 2–1.17B, and 2–1.17C.

FIGURE 2–1.17A

DEVELOPMENT OF A NARROW CONE OF LIGHT

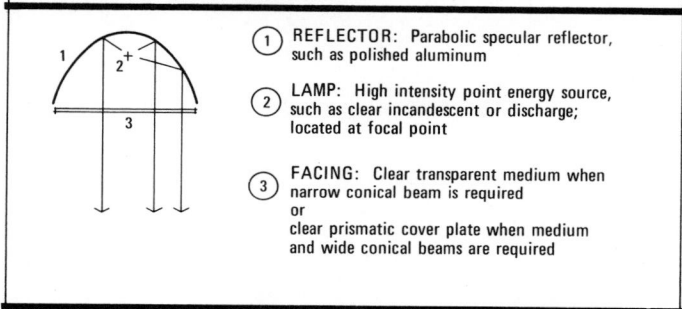

(1) REFLECTOR: Parabolic specular reflector, such as polished aluminum

(2) LAMP: High intensity point energy source, such as clear incandescent or discharge; located at focal point

(3) FACING: Clear transparent medium when narrow conical beam is required
or
clear prismatic cover plate when medium and wide conical beams are required

FIGURE 2–1.17B

DEVELOPMENT OF A DIRECTIONAL WALL WASH

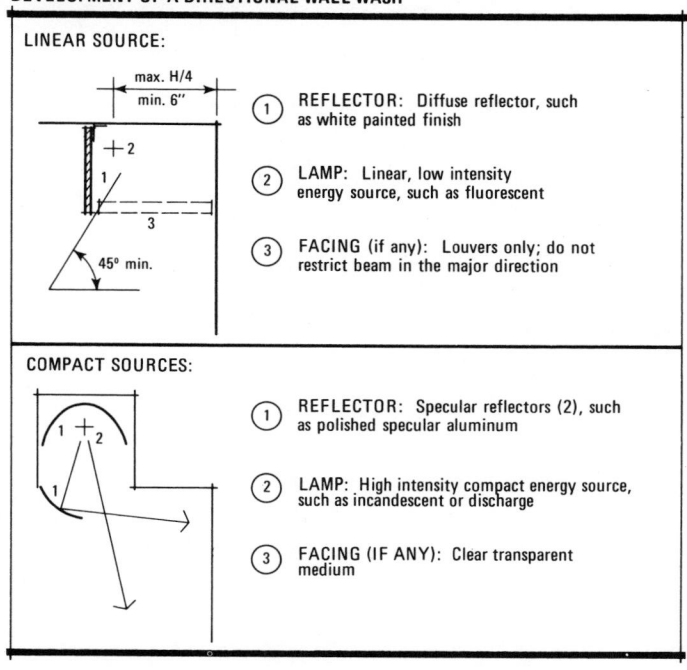

LINEAR SOURCE:

max. H/4
min. 6"

(1) REFLECTOR: Diffuse reflector, such as white painted finish

(2) LAMP: Linear, low intensity energy source, such as fluorescent

(3) FACING (if any): Louvers only; do not restrict beam in the major direction

45° min.

COMPACT SOURCES:

(1) REFLECTOR: Specular reflectors (2), such as polished specular aluminum

(2) LAMP: High intensity compact energy source, such as incandescent or discharge

(3) FACING (IF ANY): Clear transparent medium

FIGURE 2–1.17C

DEVELOPMENT OF A BROAD, DIFFUSED BEAM

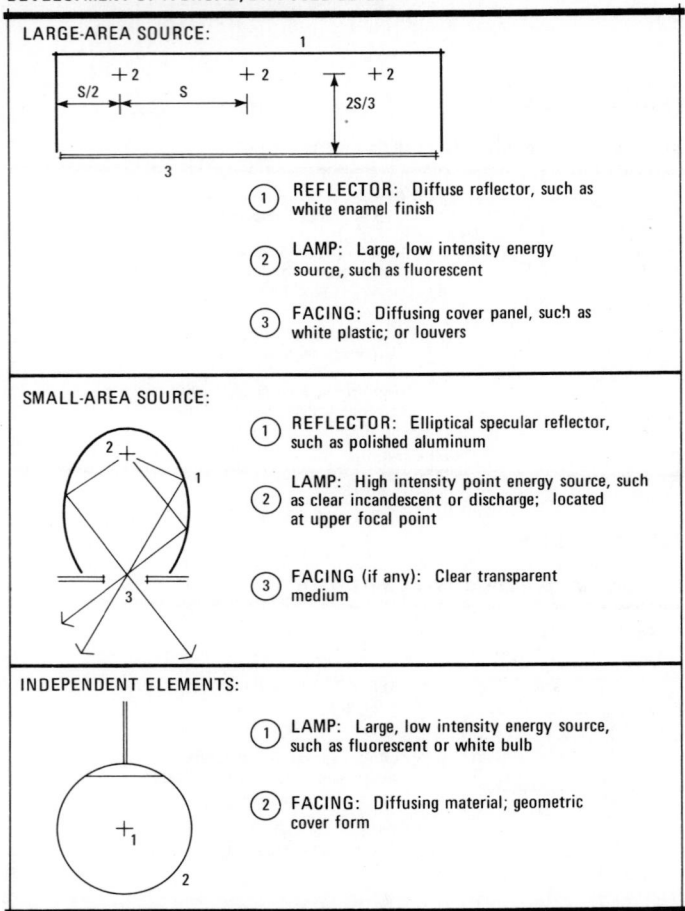

LARGE-AREA SOURCE:

① REFLECTOR: Diffuse reflector, such as white enamel finish

② LAMP: Large, low intensity energy source, such as fluorescent

③ FACING: Diffusing cover panel, such as white plastic; or louvers

SMALL-AREA SOURCE:

① REFLECTOR: Elliptical specular reflector, such as polished aluminum

② LAMP: High intensity point energy source, such as clear incandescent or discharge; located at upper focal point

③ FACING (if any): Clear transparent medium

INDEPENDENT ELEMENTS:

① LAMP: Large, low intensity energy source, such as fluorescent or white bulb

② FACING: Diffusing material; geometric cover form

Spatial influences

In the design and manipulation of the luminous environment, the designer must develop a specification which correlates two major contributing components: (1) the light emitted directly from the source toward the object or surface to be lighted, and (2) the light that is incident on the object or surface as a result of reflection from other surfaces in the space.

Direct source emission Light from a typical point source is emitted in all directions from the source. But most of this light is invisible in the sense that we *see* only that portion that is emitted toward and enters the eye. If we assume a complete spatial void, then, the eye perceives no light except the flux that flows from the source toward the eye; and the remaining light is subjectively *lost* or dissipated in the void.

If we introduce a simple diffuse reflecting sphere into the space (see Figure 2–1.18, left), a portion of the previously *lost* light intersects this form, and part of this light is reflected toward the eye. The eye now perceives a partially lighted form; and since the lower portion of the sphere receives no direct light, this aspect of the form tends to be imperceptible to the eye.

FIGURE 2–1.18

DIRECT AND SECONDARY LIGHT EMISSION

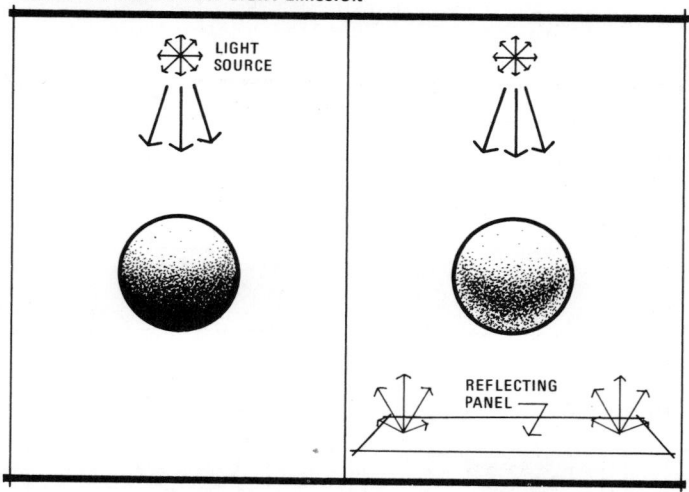

At this point, the designer can modify the perceived *color* of the sphere by several methods:

(1) He can introduce a colorant into the finish of the sphere (paint, dye, etc.). This colorant serves as a subtractive filter that absorbs certain wavelengths while reflecting others.

(2) He can introduce a filter at the light source (colored glass or plastic). This also serves as a subtractive filter, absorbing certain wavelengths and transmitting others toward the sphere.

(3) He can modify the color additively by introducing additional light sources (additional lamps, phosphors, etc.). This action shifts the spectral characteristics of the initial source by adding energy at specific wavelengths.

Secondary sources If the designer intends that the sphere be more completely illuminated in order to facilitate perception of the entire form, he must introduce a second light source (either primary or secondary) to light the lower portion of the sphere. Figure 2–1.18 (right) suggests the action of a diffuse reflecting plane, which becomes a secondary light source for redirecting light into the shadow area.

To some extent, the designer can affect the intensity of the *fill light* by changing the reflectance characteristics of this surface (high reflectance white versus low reflectance gray).

FIGURE 2–1.19

DIRECT AND INTER-REFLECTED LIGHT

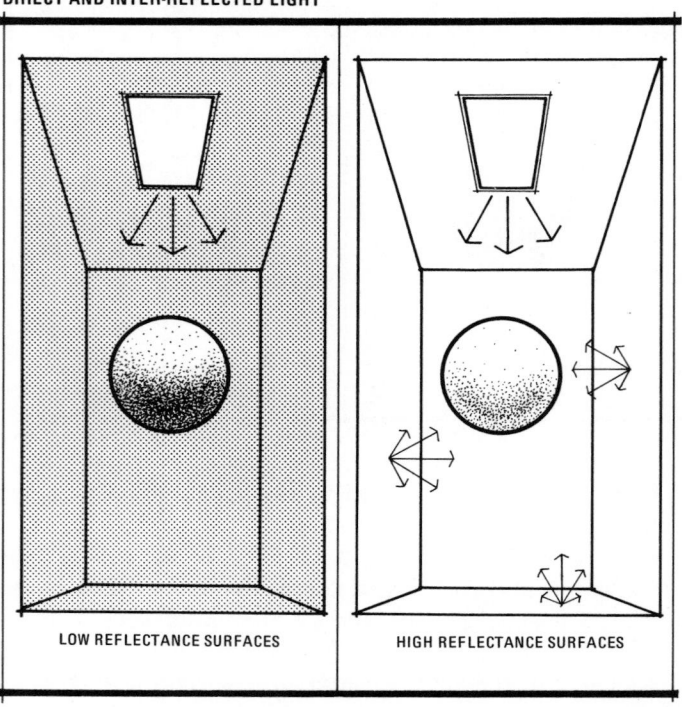

LOW REFLECTANCE SURFACES   HIGH REFLECTANCE SURFACES

He can also affect the color of the *fill light* by using the reflector surface as a subtractive filter. For example, a red surface will subtract blue and green wavelengths and reflect predominantly red light toward the shadow portion of the sphere.

Manipulation of visual space Similar relationships are developed in the analysis of light in conventional activity-oriented spaces. For example, the illumination of object form by a single direct light source may again be perceived as intense highlight and deep shadow voids; while the development of additional primary or secondary light sources will modify this perception.

The action of diffuse reflecting wall surfaces is, of course, a common influence (see Figure 2–1.19, right). In this sense, the designer can modify the intensity of interreflection (fill light) by changing the reflectance characteristics of various wall, ceiling, and floor surfaces. When all major architectural surfaces have high reflectance characteristics (near white), the secondary lighting component will become quite significant and will tend to minimize contrast by filling in the shadows.

Low reflectance enclosure surfaces (approaching black) will, on the other hand, minimize interreflection. When narrow beam spotlighting is used in this latter setting, the system will tend to produce highlighted forms within a basic *void* space.

Color of reflected light, as well as intensity, can be modified by manipulating the enclosure surface finishes.

Figures 2–1.20A and 2–1.20B suggest a further example of selective light planning in an occupied room. A highly controlled pattern of parabolic downlights (*columnated light*) is positioned to light the floor plane without impinging on the wall (note the wall-floor juncture in Figure 2–1.20A). In this situation, note that the floor is the dominant light source for lighting the ceiling. Ceiling brightness can therefore be manipulated by adjusting the reflectance of floor and table surfaces.

In Figure 2–1.20B, narrow beam units are used to light the walls with minimum effect on the ceiling and floor. In combination, then, these figures illustrate the principle of selective lighting layout which permits individual room surfaces to be manipulated in a manner somewhat independent of each other.

Figures 2–1.21 and 2–1.22 are examples of the application of this principle. In Figure 2–1.21, the horizontal plane is lighted, while the vertical field is de-emphasized (except for significant focal centers). As a fundamental contrast in spatial character, Figure 2–1.22 shows the effect of specific lighting of the vertical surfaces in a large-scale space.

FIGURE 2–1.20A
SELECTIVE CONTROL OF FLOOR AND CEILING BRIGHTNESSES

FIGURE 2–1.20B
SELECTIVE CONTROL OF WALL BRIGHTNESS

FIGURE 2–1.21

SELECTIVE SPATIAL DEVELOPMENT (SUBORDINATION OF WALLS)

FIGURE 2–1.22

SELECTIVE SPATIAL DEVELOPMENT (EMPHASIS OF WALLS)

Quantitative Estimates: Localized Distribution of Light

One approach to lighting design is highly selective. In these instances, the illumination intensity (measured in footcandles) may be high on various specific areas, while adjacent areas and surfaces tend to assume a degree of relative shadow and contrast. It is this contrast that enables the designer to be spatially selective—to choose specific surfaces or objects for special lighting treatment, while subordinating others by leaving them in relative darkness. (Also see Chapter 1–1 for discussions of *vector influences* and *spatial order and form.*)

Much of this analysis may be intuitive, as is implied in the immediate preceding discussion. However, some quantitative analysis will likely be required.

Point-by-point estimates

Point-by-point analysis at selected locations is the most effective simple means for quantitative approximations of spatial lighting effects.

Basically, this technique recognizes that the intensity of light will diminish in proportion to the square of the distance from the source to the analysis point. When the source intensity is known, then, this principle can be utilized to estimate direct light intensity (in footcandles) at any given localized point.

When this technique is to be utilized, photometric curves (available from the manufacturer) are used to determine the candlepower emission from the luminaire in the angular direction of the analysis point (see Figure 2–1.23). The following formulas then apply:

$$E_i = \frac{CP_o \times \cos I}{D^2}$$

where: E_i = initial illumination at the analysis point

CP_o = initial candlepower intensity of the source in the direction of the analysis point

$\cos I$ = cosine of the *angle of incidence* (i.e., the angular displacement from perpendicular incidence)—see Table 2–1.9

D = distance from the light source to the analysis point (in feet)

$$B_i = E_i \times R_o$$

where: B_i = approximate initial brightness (foot-lamberts)

R_o = reflectance of the analysis surface (%)

FIGURE 2–1.23

POINT-BY-POINT USE OF THE PHOTOMETRIC CURVE

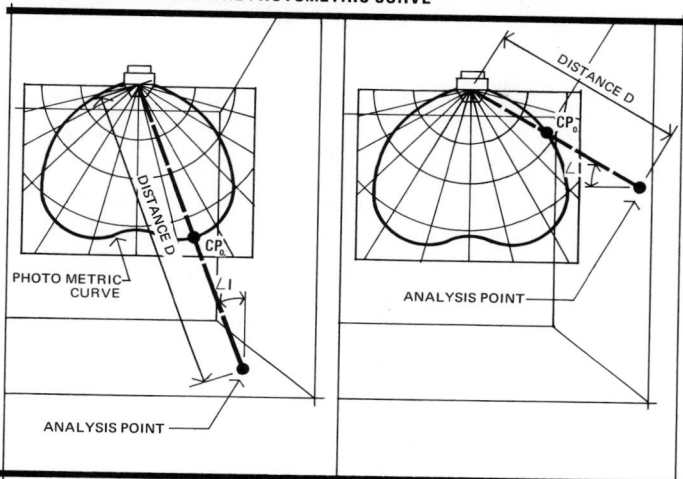

Table 2-1.9 Table of Cosines

Angle of Incidence	Cosine	Angle of Incidence	Cosine
0°	1.000	50°	0.643
5°	0.996	55°	0.574
10°	0.985	60°	0.500
15°	0.966	65°	0.423
20°	0.940	70°	0.342
25°	0.906	75°	0.259
30°	0.866	80°	0.174
35°	0.819	85°	0.087
40°	0.766	90°	0.000
45°	0.707		

Supplementary estimation of interreflection and *fill light* involves more complex calculations. Because of this lack of simple procedure, judgment of this component is often based on empirical tests, mock-up data, or intuitive estimates which utilize knowledge of the reflectance characteristics of nearby surfaces.

Quantitative Estimates: General Distribution of Light

Lighting objectives are generally oriented in one or both of two ways: (1) toward the visual needs immediately associated with a seeing or work task, and (2) toward the spatial characteristics of the visual environment. (See Chapter 1–1 for further discussion of these objectives.) In some cases, these lighting functions may be achieved selectively with separate localized systems; while in other situations, both *spatial* and *work surface* illumination is provided by the same basic system.

Where both illumination functions are to be provided by the same system, this generally involves the use of *wide beam* or *multidirectional* luminaires together with high reflectance surface finishes. This combination will diffuse light throughout the room, emitting light toward the ceiling and walls as well as toward the floor or work plane. These systems tend to produce a bright, generally uniform lighting condition; and because of the simplicity of having one lighting system provide for both functions, this design approach is often applied when broadly flexible interiors are required. Activities, furnishings, and even movable walls tend to be easily reoriented independent of the location of the lighting devices.

Design procedures are somewhat more disciplined and less intuitive than those discussed just previously.

Operational characteristics

The quantity of light produced for a given input wattage varies with the type of lamp used (see Table 2–1.6).

Both fluorescent and mercury lamps are discharge sources and produce light more efficiently than incandescence. Fluorescent lamps (which provide the designer with a wide selection of *whiteness* and color rendition characteristics) are approximately three times more efficient than incandescent lamps. They therefore require only one-third the power for a given quantity of light output.

For this reason, the fluorescent lamp is often more economical for general lighting systems, particularly in air conditioned spaces. Since these systems often lack some of the visual interest and stimulation associated with incandescent systems (i.e., sparkle, highlight, shadow, depth of color, etc.), the designer may want to introduce a supplementary incandescent system of accent lighting. The creative designer will also compensate for the *flat,* diffuse fluorescent illumination through his detailing and selection of materials, form, and color.

Quantitative performance

Quantitative illumination is usually discussed in terms of footcandles (lumens/square foot) at the work surface. For most commercial and industrial activities, this surface is a horizontal desk or table plane approximately thirty inches above the floor. (Also see Table 1–1.9 regarding recommended intensity ranges.)

Light sources are rated in terms of *lumens* (i.e., quantity of visible radiation emitted by the source). The *footcandle* is the unit of light density, representing the number of lumens that are incident on each square foot of the work surface.

In satisfying quantitative requirements with a general lighting system, then, it is necessary to develop an initial estimate of factors that reduce light intensity at the critical surface. This estimate helps the designer to compensate for expected losses and achieve a predictable approximation of light intensity.

Losses within the lighting equipment   Light is partially obstructed and absorbed by the fixture itself (by louvers, reflectors, etc.). As a result, the total lumens emitted from the fixture are less than the lumens generated by the lamp.

Losses at the room surfaces   Although some of the light from the fixture passes directly to the work surface, an additional portion may be reflected at least once from the ceiling, walls, and floor. Since no material reflects all of the light that strikes it, each reflection will involve some loss of light through absorption. It follows, therefore, that high reflectance finishes tend to reduce these losses and improve the system efficiency.

Furthermore, light reflected from a diffuse surface is scattered in many directions. When room surfaces are diffuse, and this is generally desirable, a portion of the generated light may be reflected several times before it actually reaches the work plane—with some loss of light from each reflection.

Losses due to room proportions   The greater the floor area for a given ceiling height, the more efficient is the general lighting system in delivering light to a horizontal work surface. In a large, low room, a substantial portion of the light from the fixtures reaches the work plane without reflection from the room surfaces. In a high, narrow room, on the other hand, a higher proportion of the light strikes the walls and other spatial surfaces. This latter condition results in reduced lighting efficiency relative to the horizontal work plane.

Losses due to light distribution   The distribution characteristics of the lighting equipment determine (in part) the amount of the initially-emitted light that will be reflected from various room surfaces. If most of the light is directed downward, for example, a major portion will reach the work plane with minimum absorption losses at the room surfaces. At the other extreme, with a total up-lighting system, all of the light is reflected (and partially absorbed) by room surfaces at least once.

The coefficient of utilization   The effects of these initial losses can be summarized in a single design factor, the *coefficient of utilization*. This term refers to the percentage of generated lamp lumens that actually reaches the horizontal work plane; the remainder being absorbed by the system or by other surfaces in the room. Thus a coefficient of utilization of 0.65 means that 65% of the lamp lumens are useful in producing light on the work surface.

Making use of this coefficient, the formula for estimating the effect of general lighting systems in providing initial illumination on the horizontal work plane is:

$$E_i = \frac{LL \times CU}{A_w}$$

where: E_i = average initial illumination (in footcandles or lumens/ square foot)

LL = initial lamp lumens in the system (total for the room)

CU = Coefficient of utilization (see Figure 2–1.24)

A_w = area of work plane (in square feet); for general systems, this is the area of the room

Figure 2–1.24 summarizes typical coefficient ranges for common luminaire types. This data is based on the assumption that the following limits are not violated:

	maximum spacing
pattern of fluorescent luminaires	0.9 × mounting height
pattern of downlights	0.5 × mounting height
pattern of spheres	1.0 × mounting height

	minimum distance from source-to-ceiling
up-light system	1/6 of the distance between sources

Maintenance of light output   Because initial lighting estimates are based on the lumen output of new lamps used with clean fixtures and reflecting surfaces, some depreciation should be expected over the life of the system. Depending on the cleanliness of the environment and the

FIGURE 2–1.24

TYPICAL COEFFICIENTS OF UTILIZATION

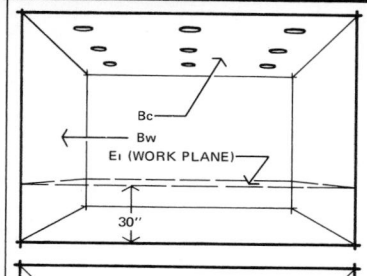

INCANDESCENT DOWNLIGHT PATTERN		
ROOM TYPE	HIGH REFLECTANCE ROOM FINISHES	LOW REFLECTANCE ROOM FINISHES
TYP. SMALLER ROOMS (moderately low ceilings)	0.58 - 0.68	0.54 - 0.63
TYP. LARGER ROOMS		
RELATIVELY HIGH CEIL.	0.60 - 0.66	0.56 - 0.63
RELATIVELY LOW CEIL.	0.65 - 0.70	0.61 - 0.65

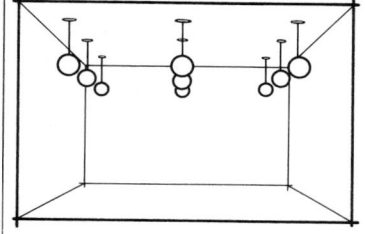

INCANDESCENT PATTERN OF SPHERES		
ROOM TYPE	HIGH REFLECTANCE ROOM FINISHES	LOW REFLECTANCE ROOM FINISHES
TYP. SMALLER ROOMS (moderately low ceilings)	0.37 - 0.57	0.25 - 0.45
TYP. LARGER ROOMS		
RELATIVELY HIGH CEIL.	0.43 - 0.61	0.30 - 0.50
RELATIVELY LOW CEIL.	0.57 - 0.68	0.45 - 0.61

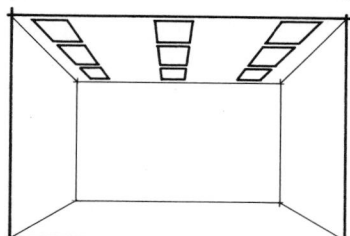

FLUORESCENT PATTERN OF 2 x 4 LUMINAIRES		
ROOM TYPE	HIGH REFLECTANCE ROOM FINISHES	LOW REFLECTANCE ROOM FINISHES
TYP. SMALLER ROOMS (moderately low ceilings)	0.32 - 0.49	0.22 - 0.40
TYP. LARGER ROOMS		
RELATIVELY HIGH CEIL.	0.37 - 0.52	0.27 - 0.43
RELATIVELY LOW CEIL.	0.49 - 0.58	0.40 - 0.53

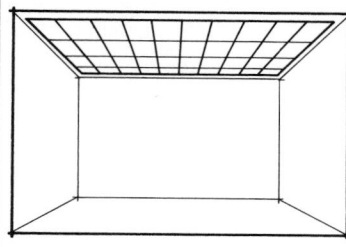

FLUORESCENT LUMINOUS CEILING		
ROOM TYPE	HIGH REFLECTANCE ROOM FINISHES	LOW REFLECTANCE ROOM FINISHES
TYP. SMALLER ROOMS (moderately low ceilings)	0.23 - 0.32	0.19 - 0.29
TYP. LARGER ROOMS		
RELATIVELY HIGH CEIL.	0.25 - 0.33	0.21 - 0.30
RELATIVELY LOW CEIL.	0.32 - 0.36	0.29 - 0.34

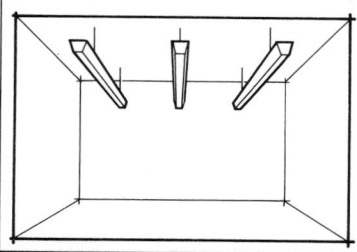

FLUORESCENT UPLIGHT		
ROOM TYPE	HIGH REFLECTANCE ROOM FINISHES	LOW REFLECTANCE ROOM FINISHES
NORMALLY NOT USED IN LOW CEILING ROOMS		
TYP. LARGER ROOMS		
RELATIVELY HIGH CEIL.	0.18 - 0.33	0.07 - 0.17
RELATIVELY LOW CEIL.	0.29 - 0.39	0.14 - 0.23

cleaning and lamp replacement schedule (as well as the type of lamp and fixture), a design allowance is usually made to compensate for the expected reduction in light output.

An assumed reduction to 70–80% is typical for clean or air conditioned spaces; and an assumed reduction to 60–70% is typical for more adverse conditions. This design estimate is referred to as the *maintenance factor.*

Using this factor, the formula for estimating the illumination produced by a system after a period of service is:

$$E_m = E_i \times MF$$

where: E_m = maintained illumination in service (footcandles)
   E_i = initial illumination
   MF = maintenance factor

Estimates of spatial brightness and interreflection   It must be remembered, however, that the coefficient of utilization refers only to the effect of general lighting systems in providing light on the single *horizontal* plane. The term therefore constitutes one index of system performance relative to typical work or seeing tasks. But it makes no allowance for the spatial factors that also affect the visual environment (see Chapter 1–1).

To assist in further evaluation, the factors shown in Table 2–1.10 can be used to develop general approximations of wall and ceiling brightness for the representative general systems noted in Figure 2–1.24.

Table 2-1.10 Wall and Ceiling Factors

	Wall Factors	Ceiling Factors
Downlight pattern	0.30–0.45	0.12–0.20
Pattern of spheres	0.85–1.00	0.80–1.20
Fluorescent pattern	0.30–0.50	0.12–0.25
Luminous ceiling	0.60–0.75	——
Uplight	0.60–0.75	1.50–3.00 For high reflectance room finishes 2.50–6.00 For low reflectance room finishes

Using these factors, the formula for estimating the effect of general lighting systems in producing wall brightness is:

$$B_w = E_{wp} \times WF \times R_w$$

where: B_w = approximate average wall brightness at the floor-to-ceiling midpoint (in foot-lamberts)

E_{wp} = average horizontal work plane illumination (in footcandles)

WF = appropriate wall factor (see Table 2–1.10)

R_w = reflectance of the relevant wall (%)

Similarly, the formula for estimation of ceiling brightness is:

$$B_c = E_{wp} \times CF \times R_c$$

where: B_c = approximate average ceiling brightness (in foot-lamberts)

CF = appropriate ceiling factor (see Table 2–1.10)

R_c = reflectance of the ceiling (%)

Table 2-1.11   Recommended Reflectances for General Systems

	Reflectance Range
Ceiling finishes	70–90%
Wall finishes	40–70%
Floor finishes	10–40%
Desk and bench tops	25–50%

NOTE: Indicated reflectance ranges refer to mat (diffuse) surfaces and finishes.

Calculated Factors Relating to Luminaire Selection

When estimating the comprehensive performance of general lighting systems, Table 2–1.12 is useful for summary purposes.

Intuitive Factors Relating to Luminaire Selection

In addition to functional aspects related to the light produced in the space, luminaires and other components of the lighting system become, in themselves, potentially prominent factors in the visual composition of space. The mechanical elements of lighting design must therefore be analysed by architectural standards, as well as for their engineering function and performance (see Chapter 1–1 for discussion of "Spatial Order and Form").

In this regard, a study of architectural history reveals two basic alternatives in the approach to lighting design: (1) the *visually subordinate*

Table 2-1.12 Estimation of Lighting System Performance (Intensity)

	Formulas	References
General Illumination:		
Horizontal Illumination	$E_1 = \dfrac{LL \times CU}{A_W}$ (initial illumination)	See Figure 2–1.24
	$E_M = E_1 \times MF$ (maintained illumination)	See Table 1–1.9 for criteria
Spatial Illumination	$B_W = E_{WP} \times WF \times R_W$ (wall brightness)	See Table 2–1.10
Luminaire Brightness and Glare	$B_C = E_{WP} \times CF \times R_C$ (ceiling brightness)	See Table 1–1.2 for criteria
Supplementary Illumination:		
Spotlight Patterns	Point-by-point calculation or utilize pre-calculated graphic beam patterns	See Table 1–1.2 for criteria
Wall Lighting	Point-by-point calculation or utilize pre-calculated graphic beam patterns	See Tables 1–1.2 and 1–1.4 for criteria

system and (2) the *visually prominent system.* These alternatives are still seen in the approach of various individual architects toward lighting design.

Visually subordinate lighting systems

Some design concepts reflect an attempt to introduce light in a manner that the occupant will be conscious of the effect of the light, while the light source itself is subordinate. For example, in some Byzantine churches, small unobtrusive windows were placed at the base of a dome to light this large structural element. The brilliant dome then became a dominant

FIGURE 2–1.25

THE VISUALLY SUBORDINANT LIGHTING SYSTEM

visual factor in the space; and serving as a huge reflector, the dome (not the windows) became the apparent primary light source for the interior space. Similarly, in some Baroque interiors, the observer's attention is focused on a brightly lighted decorative wall, while an adjacent window was placed at the side so as to be somewhat concealed from the normal view of the typical observer.

In both cases, the objective was to place emphasis on the surfaces to be lighted, while minimizing any distracting influence from the light source itself.

In these historical instances, daylighting systems were involved. But this same design attitude can be seen in the development of some electric lighting systems (see figure 2–1.25). These systems utilize compact, directional lighting units and/or lighting units that use extremely low-brightness shielding devices. Such devices direct light toward a specific surface, plane, or object; emphasizing these areas with little distracting influence from the lighting unit itself. Inherently, then, the space is visually defined as a composition of *reflected* brightness patterns (horizontal and vertical).

FIGURE 2–1.26
DELINEATION OF SPATIAL FORM

Accurate delineation of spatial form and surface character

The *form* of the light distribution pattern (i.e., a *cone* of light, an *area wash* of light, etc.) should relate logically to the form of the affected surface. A wall or ceiling plane, for example, should generally be lighted with an area wash of light which approximates, as near as possible, the form and dimension of the surface involved (see right and left walls in Figure 2–1.26). *Scallops* and similar irrelevant variations should be minimized; for except when special effects are involved (such as those which develop sparkle and highlight as a means to instill a sense of vitality), the surface should be basically perceived as an integrated form, not as a form or surface intersected by meaningless patterns of light.

Similarly, the system should perform in sympathy with the surface character involved, with grazing light to complement a textured surface, or a more diffuse or frontal light for a flat surface.

FIGURE 2–1.27
THE VISUALLY PROMINENT LIGHTING SYSTEM

Visually prominent lighting systems

Particularly when broad-beam or multidirectional devices are involved, a light source or luminaire may also attract attention to itself, even to the extent that such elements become particularly dominant factors in the design. In architectural history, the large stained glass windows of the Gothic period are probably the most obvious examples of this approach to lighting design. In contemporary building, transilluminated ceilings and walls are a similar dominant influence.

In cases where light-transmitting (rather than opaque) materials are predominant in the lighting device, then, these units become architectural forms and building surfaces as well as lighting elements. In this sense, self-luminous elements also help to visually define a space, and they are important in the general spatial organization of the room (see Figure 2–1.27).

Integrity of system forms   When environmental components are specified or designed, the physical form of the devices themselves becomes a consideration. Simple geometric forms such as a sphere, a cube, a cone, a cylinder, a rectangular luminous surface, etc. are generally more enduring than contrived forms—provided that these basic shapes are appropriate technically as well as aesthetically. The luminous sphere is a widely used example; and it is to be noted that the physical form of this device also expresses the shape of the multidirectional spread of light that is produced. Similarly, the directional nature of a spotlight beam is geometrically expressed by a simple cylindrical housing.

Sound Generation and Control

Sound is initially generated by a vibrating source, such as the action of the vocal chords or the action of a musical instrument. These vibrations induce small atmospheric changes that alternately vary above and below normal atmospheric pressure[1] (see Figure 2–2.1). The average deviation in pressure is called *sound pressure* (intensity); and by causing the listener's ear drum to vibrate in sympathy, these pressure variations produce a sensation of hearing.

The tone of the signal is determined by the rate at which the pressure alternates above and below the ambient atmospheric condition. This variable is called *frequency* and is determined by the rate of vibration at the source. High rates of source vibration will produce short wavelength, high-pitched sounds; low vibration rates will produce a low-pitched or low frequency sound.

Sounds may originate within the immediate space—such as conversational speech and other human or mechanical sounds associated with an activity. They may also originate external to the primary space—such as vehicular traffic noise, railroad or aircraft noise, sounds associated with playground activity, wind and other natural sounds, and noises associated with activities in adjacent or nearby rooms.

In sonic design, then, the designer must develop a spatial specification that (1) assists in defining the meaningful sound signals required for communication and orientation, and (2) insures that these signals are perceived against an acceptable sonic background.

---

[1] Sound vibrations can also be transmitted through liquid and solid mediums. Some such transmitting medium must be present, however; as can be demonstrated by operating a bell inside of an exhaustible jar. As long as air remains in the jar, the bell is heard ringing. But when the air is withdrawn, the bell continues to operate without being heard.

149

FIGURE 2–2.1

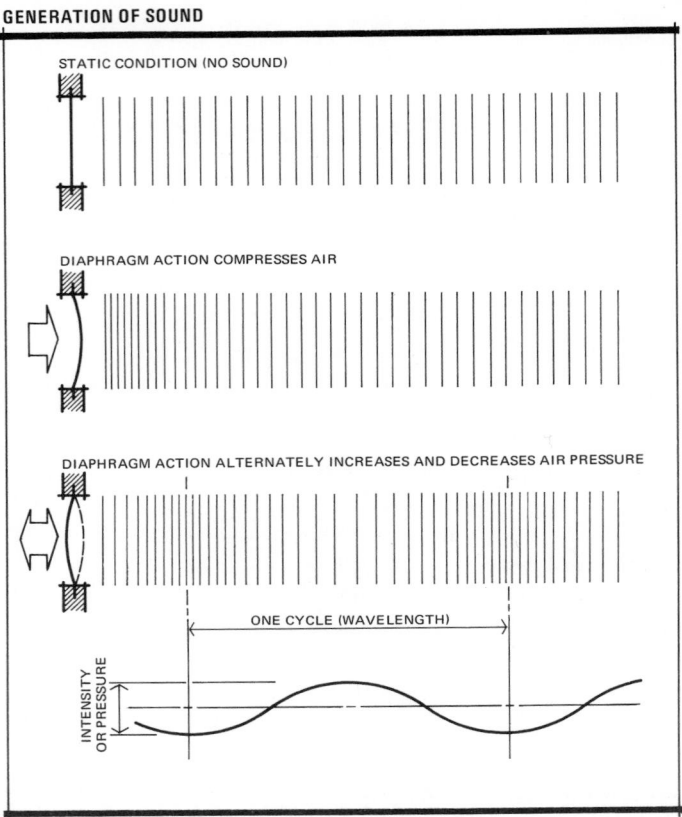

GENERATION OF SOUND

REINFORCING THE PRIMARY COMMUNICATION SIGNAL

Initially, sound will travel outward in all directions from the source in a manner somewhat comparable with the effect of water waves as they travel outward from the point at which a stone is dropped. As the wave travels, pressure (and therefore intensity) diminishes with distance.

But when the sound originates within an enclosure, the initial wave pattern may also be intersected by other waves that are reflected from the enclosing shell. At each of the enclosing surfaces, sound energy will be partially absorbed and partially reflected. The overall spatial effect of this action will significantly influence the sonic quality of the space because this determines the tendency of sound signals to be either subtractively dampened or to mix additively and reinforce (or muddle) the direct signal.

For a sustained sound in an enclosed space, then, the ear hears both *direct sound* (as diminished by distance) and *reflected sound* (as redirected and diminished by the action of room surfaces). When the source and listener are in close physical proximity, the direct component is dominant and may, in itself, be totally adequate for effective communication. But as the distances become greater, it becomes increasingly necessary for the designer to develop a method to implement or reinforce weak signals.

This reinforcement can be produced by one or both of two methods: (1) by *natural amplification* techniques that utilize surface forms and materials to selectively reflect peripheral sound energy in a manner that will reinforce the direct signal, and (2) by the use of various *electronic amplification* techniques which either increase the intensity of the initial source or transmit the signal to more remote locations.

Material Action

Reinforcement or suppression of sound depends on careful selection and placement of sound reflectors and barriers.

When a sound wave strikes a sizable surface or object, part of the energy is absorbed and the remainder is transmitted or reflected. Signal distribution within an enclosure can therefore be partially manipulated through reflection; while intensity and frequency balance can be partially manipulated through absorption.

FIGURE 2–2.2

REFLECTION OF SOUND

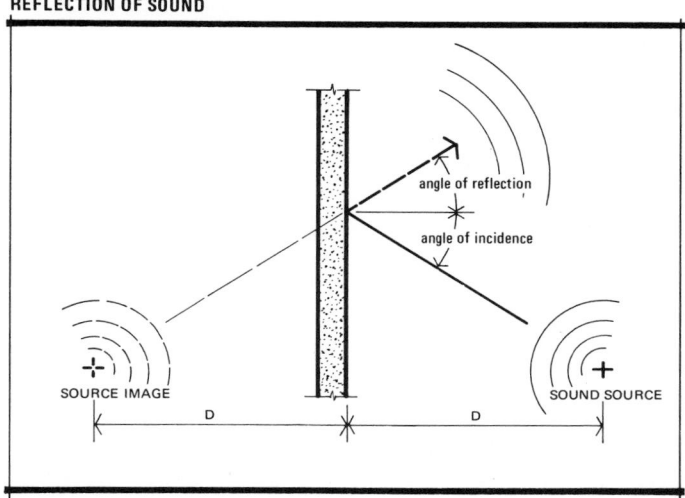

Reflection

Sound, like light, is reflected approximately according to the conventional laws of reflection (i.e., angle of incidence equals angle of reflection).

If the reflecting surface is a plane, the plotting of apparent *sound images* behind the plane is a useful technique for evaluating the characteristics of that surface in redirecting sound (see Figure 2–2.2).

Surface configuration   The form or configuration of room surfaces will therefore affect the sound reinforcing properties of the enclosure—redirecting and focusing energy, or diffusing and mixing sound.

Hard, nonporous, and *continuous* surfaces (such as large panels of plaster, wood, or concrete) are highly reflective over the entire range of relevant frequencies. Such surfaces are somewhat analogous with a *specular* reflecting surface in lighting design.

Hard, nonporous, but *irregular* surfaces (such as coffered plaster ceilings, or thin-shell concrete folded plates) are, on the other hand, analogous with a white *diffusing* material in lighting (see Figure 2–2.3). In this case, effective diffusion of sound will result from large-scale surface irregularities, normally at least 3–4 ft or more across and 6 in. or more in depth. The dimension of each facet or reflecting plane must equal or exceed the longest relevant wavelength; so small-scale configurations will only affect high frequency (short wavelength) sounds.

Concave surface curvature   Continuing the general analogy with the action of light, large unbroken concave surfaces will tend to redistribute sound energy in an irregular manner. As a result, these surfaces do not satisfy the normal requirements of uniform signal distribution.

For this reason, focusing shapes should be avoided. Or, if this is not possible, superimposed reflecting panels or large-scale surface modulation may be required (see Figure 2–2.4). Still another method is the development of curved forms as *acoustically transparent* surfaces, such as occurs with the use of separated wood strips backed by an air space that includes appropriately angled reflective panels or highly absorbent materials.

Natural Sound Reinforcement

When sound is emitted from a single source toward an audience, the sound energy received by those in the rear is reduced: (1) by normal decay due to distance, and (2) by the absorption effect of the audience in front. The Greeks and Romans minimized these losses by placing the

FIGURE 2–2.3

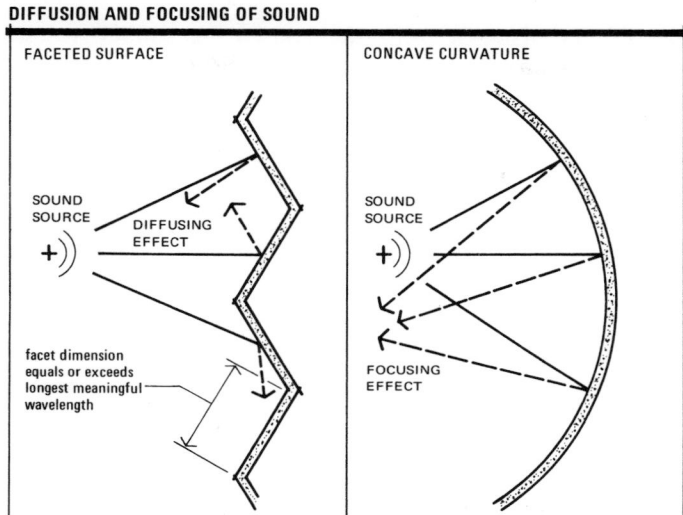

DIFFUSION AND FOCUSING OF SOUND

FACETED SURFACE

SOUND
SOURCE
DIFFUSING
EFFECT

facet dimension
equals or exceeds
longest meaningful
wavelength

CONCAVE CURVATURE

SOUND
SOURCE

FOCUSING
EFFECT

FIGURE 2–2.4
LARGE SCALE DIFFUSER PANELS IN A CONCAVE INTERIOR

audience on a steep slope in an outdoor theater. This action minimizes the effect of audience absorption and makes signal intensity almost directly dependent on the distance from the source. A similar result can be achieved by placing the sound source high above the audience, but this may cause an uncomfortable or inappropriate visual relationship.

This latter effect is simulated in interior spaces when a hard reflective ceiling is utilized as a *sound mirror* (see Figure 2–2.5). When high intensity signal distribution is required, this mirror can be shaped and refined in a number of ways in order to maximize the area of reflective surface that individual listeners will *see* from their listening positions in the audience.

When differing sound signals are emitted concurrently by a large group, such as an orchestra or chorus, the sound-reflective surface serves to

FIGURE 2–2.5

THE REINFORCING SOUND 'MIRROR'

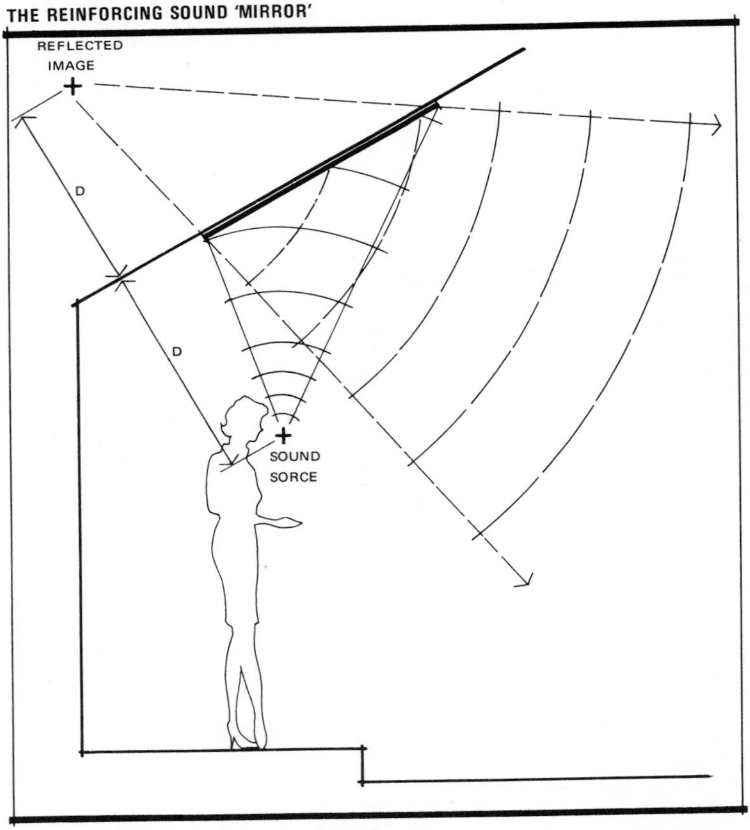

diffuse and mix the sound. Diffusion is important in providing consistent communication with the audience—to insure that various listeners are hearing the same thing. Diffusion is also important for the performers themselves, for it is essential that all performers hear each other. This latter need generally leads to the use of overhead or vertical faceted reflectors in the immediate vicinity of the performing group. This may take the form of coffers or folded plates.

Essentially, then, design for natural amplification in an enclosed space leads toward the use of surfaces (1) that are hard and reflective; (2) that

FIGURE 2–2.6

RAY DIAGRAMMING

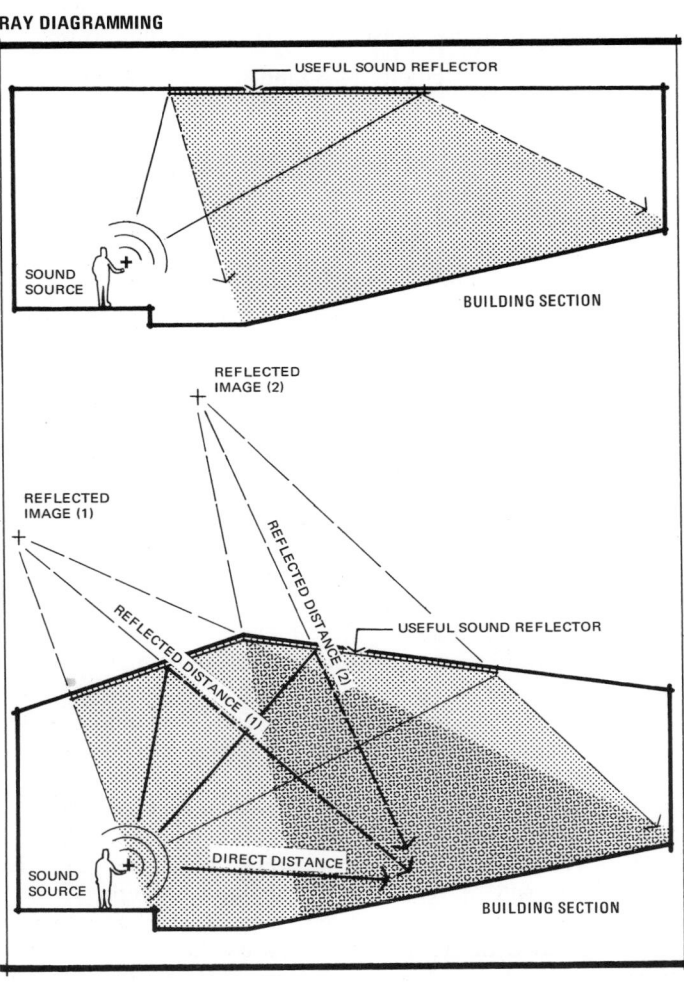

redirect sound from the source to the listener; and (3) that provide moderate diffusion to insure adequate mixing and uniformity over the entire audience.

Ray diagrams

Reflected sound in an enclosed space can be studied by a method called *ray diagramming* (see Figure 2–2.6). This involves an analysis which recognizes and utilizes the previously discussed sound reflecting characteristics of major room surfaces.

The smallest dimension of an individual reflector facet or panel must be greater than the wavelength of the lowest significant frequency. In this respect, the following formula is useful:

$$\text{Wavelength (feet)} = \frac{\text{Velocity of Sound}}{\text{Frequency}} = \frac{1140 \text{ ft/sec}}{\text{(cycles per second)}}$$

This minimum dimension requirement generally leads to panel dimensions of 4 ft or more in a space that is intended to facilitate improved verbal communication. Music may require slightly larger panels because lower frequency sounds are significant.

These panels or facets must be organized within a spatial *volume* that is determined by *reverberation time* requirements (discussed later in this chapter).

Table 2-2.1 Ray Diagramming and Natural Amplification Systems

NOTE: Refer to Figure 2-2.6

Objective 1:	To develop ceiling and wall forms to maximize the useful reflector surface
Objective 2:	To avoid *standing waves,* where sound is reflected back and forth between two parallel surfaces of approximately equal dimensions
Objective 3:	To avoid focusing effects, such as those caused by concave curvature
Objective 4:	To provide for blending of sound and diffusion through faceting of major surfaces (faceting provides overlapping of reflected sound images)
Objective 5:	To arrange useful facets so that first reflection sounds arrive 0.035 seconds or less after the direct signal
Objective 6:	To manipulate remote materials and forms in a manner that will prevent echo or *muddling* conditions
Objective 7:	To properly manipulate materials and space volume to provide the proper reverberation time

When projection of speech is important, it is desirable to arrange the reflecting surfaces so that the maximum first reflection sounds arrive at each seating location within 0.035 sec after the direct signal. This means that the reflected path should generally not exceed the direct path by more than 40 ft. If this dimensional limit is exceeded, the reflected signals may tend to become a muddling or blurring influence at the listening location in question. Furthermore, when the reflected path is more than 65 or 70 ft longer than the direct path, a distant echo will be possible. In both of these latter situations, the offending surface(s) should be treated with absorbing materials or otherwise angled or configured to *ground* the conflicting sound reflections.

Estimating the perception of loudness

The decibel scale is logarithmic as a measure of power level. A *0* db sound at the base frequency of 1000 cps is barely audible; a *10* db sound is 10 times as intense; a *20* db sound is 10 times the intensity of the 10 db sound (i.e., 100 times the intensity of a 0 db sound); a *30* db sound is 10 times the intensity of a 20 db sound; and so on.

This scale also provides a method for estimating the *subjective* effect of sound contrast. A listener interprets a 10 db increase as approximately twice as loud; a 20 db increase is perceived as approximately four times as loud; etc.

As a method for interpretation of both power level and subjective interpretation, then, this scale is particularly useful for quick estimates (see Table 2–2.2). For example, intensity tends to diminish with the square of the distance. Since 6 db approximates a four-fold change in power level, then, the designer can anticipate a drop of approximately 6 db for every doubling of the distance from the source to the receiver. The subjective sense of change or contrast for a given listener location can then be interpreted from the *apparent change* column of Table 2–2.2.

Additive sonic signals   In ray diagramming, the direct signal (diminished by distance) and the reflected signals (diminished by distance and surface absorption) are approximately additive if the arrival times are less than 0.035 sec apart. However, since decibels are logarithmic units, they cannot be added directly. Figure 2–2.7 is provided to assist in estimating additive intensity.

Electronic Amplification

As a general rule, seating capacities of 600 or less should be capable of adequate signal reinforcement by natural means; while occupant capacities

Table 2-2.2 Subjective Perception of Changes in Loudness

Change in Measured Intensity	Apparent Change (Subjective Sense of Change)	Actual Change in Power Level
1 Decibel	Base unit of measurement	1.25 (or 0.8) times base power level
3 Decibels	Barely perceptible change	2 (or 0.5) times base power level
6 Decibels	Change perceptible with doubling (or halving) the distance from the source	4 (or 0.25) times base power level
7 Decibels	Clearly perceptible change	5 (or 0.2) times base power level
10 Decibels	Change perceptible as *twice as loud* (or *half as loud*)	10 (or 0.1) times base power level
20 Decibels	Change perceptible as *four times as loud* (or *one-fourth as loud*)	100 (or 0.01) times base power level
30 Decibels	Change perceptible as *eight times as loud* (or *one-eighth as loud*)	1000 (or 0.001) times base power level
40 Decibels	Change perceptible as *sixteen times as loud* (or *one-sixteenth as loud*)	10^4 (or 10^{-4}) times base power level
100 Decibels	Change perceptible as *1000 times as loud* (or *0.001 times as loud*)	10^{10} (or 10^{-10}) times base power level

of 1000 or more will nearly always require some electronic amplification. Electronic systems will probably also be required for sound projection in large, low-ceiling, and *soft* rooms.

FIGURE 2–2.7

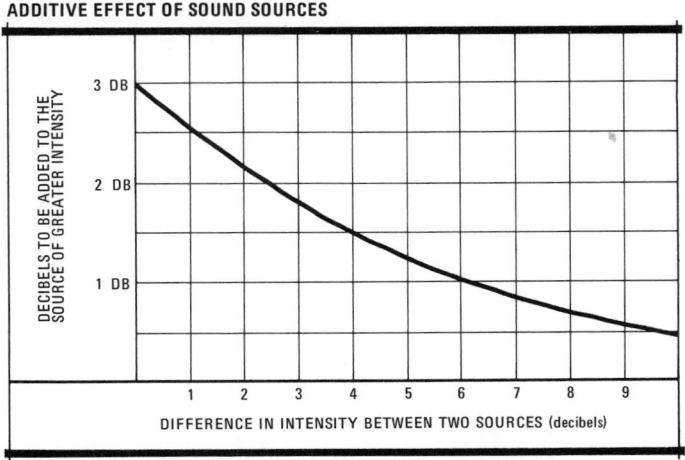

ADDITIVE EFFECT OF SOUND SOURCES

DIFFERENCE IN INTENSITY BETWEEN TWO SOURCES (decibels)

Generally, a single *central* speaker (or speaker cluster) located near the visual source is preferable over a distributed speaker system. This central approach will produce an amplified sound with the same basic directional and time characteristics as the original emitted sound.

To prevent *feedback,* which is caused by amplified sound being picked up by the microphone, large highly directional speakers should be used (typical size: 8 × 8 × 3 ft). This approach is sometimes described as a *high level system.*

In low-ceiling rooms, however, a central speaker system may not be practical. A distributed system of small individual speakers (typically 8 in) should be mounted so they project sound directly down from the ceiling. This approach is sometimes described as a *low level system.*

In either case, care should be exercised to prevent a late arrival of *live* direct sound. This lag behind amplified sound can produce muddling or echoes. Correction can be provided either by minimizing the intensity of the *live* sound or by introducing a slight time lag in the amplification.

Generally, a listener without visual contact with the source, will associate the location of the sound source with the first sound that reaches him, even if subsequent arrivals are more intense. This precedence effect must be considered in adjusting time lags in the amplification system.

CONTROL OF INTERNALLY-GENERATED NOISE

Sonic communication is facilitated by a high signal to background ratio. In this sense, natural or electronic amplification is useful for reinforcing the primary signal itself. But in many cases, improved control of background noise is also logical and effective as a supplementary (or primary) spatial treatment.

Material Action

Irrelevant signals or noise can be partially suppressed through the subtractive action of surface absorption. This action produces a decrease in the pressure of a reflected sound, relative to the pressure or intensity of the incident signal.

Speed of energy travel is a constant and is not affected by absorption.

Absorption

Changes in the directional characteristics of a sound wave do not reduce its intensity level, except to the extent that some of the energy is absorbed by the surfaces which it strikes. The proportion of incident sound pressure that is absorbed during a given reflection is called the *sound absorption coefficient* of that particular reflecting surface. For example, hard, nonporous surfaces such as plaster, glass, wood, concrete, and most sheet plastics generally have low absorption coefficients of 0.05 or less (i.e., approximately 95% or more of the incident energy is reflected or transmitted).

When the designer needs a material to reduce or dampen the internal noise level, then, he is looking, not for a hard, nonporous material, but for a porous or soft material that will permit sound waves to penetrate. This penetration causes sound to lose energy by frictional drag before it re-enters the room; so much higher absorption coefficients are involved. Acoustical blankets and acoustic tile are examples of materials that exhibit a high absorption coefficient, as are carpeting, heavy fabrics, clothing, and upholstery.

Panel flexure by thin plywood or plastic sheeting may also produce some dampening of low frequency sounds. But this characteristic can be utilized for noise reduction only when the flexing panel is backed by a relatively large unoccupied space (such as a lighting cavity). This is a somewhat unique case, however, and noise reduction is more commonly produced *within* the absorbing material rather than by surface flexure.

FIGURE 2–2.8

ABSORBING BAFFLES

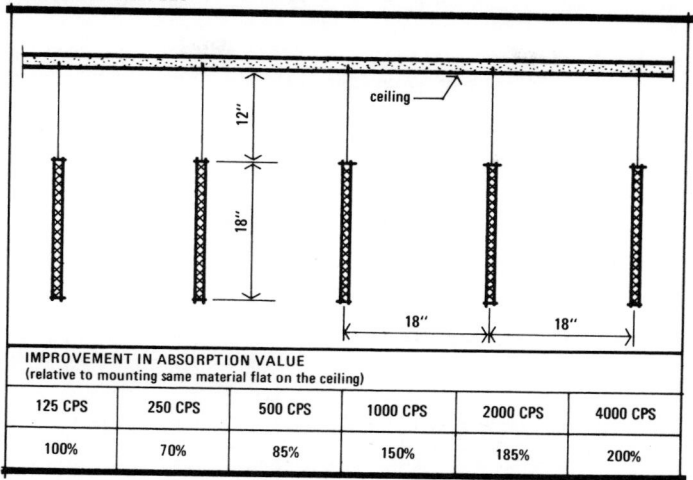

IMPROVEMENT IN ABSORPTION VALUE (relative to mounting same material flat on the ceiling)					
125 CPS	250 CPS	500 CPS	1000 CPS	2000 CPS	4000 CPS
100%	70%	85%	150%	185%	200%

FIGURE 2–2.9

INSTALLATION AND FINISH VARIABLES

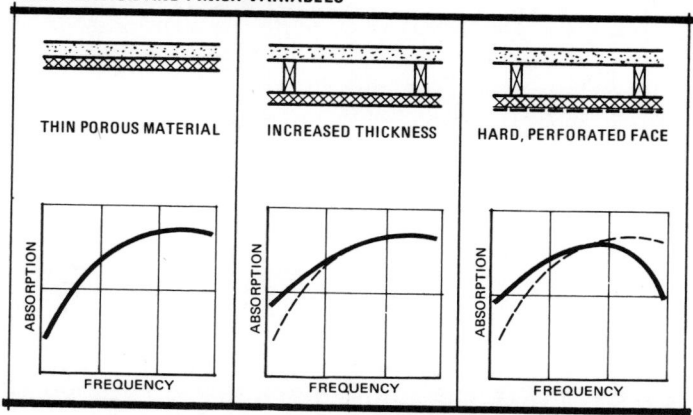

Characteristics of absorbing materials The absorption qualities of porous materials are basically determined by material thickness and surface finish (see Figure 2–2.9).

(1) *Backing:* When porous materials are mounted directly on a hard surface (such as concrete, plaster, or gypsum board), the system produces good absorption of middle and high frequencies. However, low frequencies are absorbed very little.

Table 2-2.3 Absorption Coefficients

	ABSORPTION COEFFICIENTS					
	125 cps	250 cps	500 cps	1000 cps	2000 cps	4000 cps
TYPICAL WALL MATERIALS						
High reflectance:						
Unglazed brick	0.03	0.03	0.03	0.04	0.05	0.07
Plaster on brick	0.13	0.15	0.02	0.03	0.04	0.05
Marble or glazed tile	0.01	0.01	0.01	0.01	0.02	0.02
Concrete block, unpainted	0.36	0.44	0.31	0.29	0.39	0.25
Concrete block, painted	0.10	0.05	0.06	0.07	0.09	0.08
Plaster on lath	0.02	0.03	0.04	0.05	0.04	0.03
½-in gypsum board panelling	0.29	0.10	0.05	0.04	0.07	0.09
⅜-in plywood panelling	0.28	0.22	0.17	0.09	0.10	0.11
Glass, typical window	0.35	0.25	0.18	0.12	0.07	0.04
Glass, heavy plate	0.18	0.06	0.04	0.03	0.02	0.02
High absorption:						
Lightweight drapery, 10 oz/sq yd (flat on wall)	0.03	0.04	0.11	0.17	0.24	0.35
Mediumweight drapery, 14 oz/sq yd (drape to half area)	0.07	0.31	0.49	0.75	0.70	0.60
Heavyweight drapery, 18 oz/sq yd (drape to half area)	0.14	0.35	0.55	0.72	0.70	0.65
TYPICAL FLOOR MATERIALS						
High reflectance:						
Marble or glazed tile	0.01	0.01	0.01	0.01	0.02	0.02
Concrete	0.01	0.01	0.015	0.02	0.02	0.02
Tile on concrete	0.02	0.03	0.03	0.03	0.03	0.02
Wood	0.15	0.11	0.10	0.07	0.06	0.07

High absorption:						
Heavy carpet, on concrete	0.02	0.06	0.14	0.37	0.60	0.65
Heavy carpet, on padding	0.08	0.24	0.57	0.69	0.71	0.73
TYPICAL CEILING MATERIALS						
High reflectance:						
Concrete	0.01	0.01	0.015	0.02	0.02	0.02
Plaster on lath	0.02	0.03	0.04	0.05	0.04	0.03
½-in gypsum board panelling	0.29	0.10	0.05	0.04	0.07	0.09
⅜-in plywood panelling	0.28	0.22	0.17	0.09	0.10	0.11
High absorption:						
½-in fissured tile (24 × 48)	0.33	0.39	0.53	0.77	0.86	0.80
¾-in fissured tile (12 × 12)	0.44	0.44	0.59	0.78	0.84	0.79
½-in textured tile (12 × 12)	0.11	0.23	0.65	0.82	0.80	0.74
⅝-in fire-resistive, ventilating tile (24 × 48)	0.45	0.45	0.59	0.82	0.83	0.65
TYPICAL FURNISHINGS AND OCCUPANCY CONDITIONS						
Leather seating, unoccupied	0.44	0.54	0.60	0.62	0.58	0.50
Upholstered seating, unoccupied	0.49	0.66	0.80	0.88	0.82	0.70
Upholstered seating, occupied	0.60	0.74	0.88	0.96	0.93	0.85
EFFECT OF OPENINGS						
(Effect will vary depending on the absorption or volume on the other side of the opening)						
Large openings	—		0.50–1.00	—		
Small openings	—		0.15–0.50	—		
Absorption effects of air (Sabins/1000 cu ft)	—	—	—	—	2.3	7.2

(2) *thickness:* Low frequency absorption is improved when the thickness of the absorbing material is increased. This can be done either by providing a thicker porous blanket or by providing an air space behind the material.

(3) *facing:* A hard, perforated facing or finish will reduce absorption of high frequency sounds. This varies somewhat with the size and spacing of openings (holes) in the facing.

Very thin plastic, paint, or paper films can be used as protective coverings for maintenance purposes. In these cases, relatively little energy is reflected, and most of the incident sound is transmitted directly to the porous core.

Room Absorption

The term *room absorption* is an expression of the total absorption influences within a room.

To obtain this quantity, the area of each room surface is multiplied by its absorption coefficient (for example, 100 sq ft $\times$ 0.60 = a surface absorption of 60 sabins).[2] The sum of these surface effects is added to the absorption of significant individual objects and furnishings in order to derive a total *room absorption*. In large spaces, there may be a slight addition due to the effect of the air mass itself.

This relationship is summarized in the following formula:

$$\Sigma SA = S_1 A_1 + S_2 A_2 + S_3 A_3 + \ldots$$

where:

$$\Sigma SA = \text{total room absorption (in sabins)}$$
$$S_1, S_2, S_3, \text{ etc.} = \text{area of each surface or object}$$
$$A_1, A_2, A_3, \text{ etc.} = \text{absorption coefficient of each surface or object (see Table 2–2.3)}$$

In critical situations, such calculations should be developed for each frequency band. However, as a general indication of room performance, the 500 cps band is usually evaluated as typical.

The appropriateness of various room absorption totals can be interpreted with the assistance of Figures 1–2.4 and 1–2.8.

Estimating noise control value (internal noise)

Multiple room reflections tend to increase the general noise level in active spaces and obscure the clarity, location, and identity of meaningful

---

[2] One sabin is the absorption value of one sq ft of surface that has theoretically perfect absorption qualities (i.e., an absorption coefficient of 1.0). Therefore, the above 100 sq ft example which provides an absorption of 60 sabins is the equivalent of a 60 sq ft surface which has an idealized absorption coefficient of 1.0.

signals. For most *general purpose spaces,* then, it is desirable to dampen reflected signals in order to reduce the background intensity below that required for easy conversation.

The intensity of reflected noise varies inversely with the room absorption. In this sense, the precise spatial effect or value of specific material changes can be estimated with the following formula:

$$NR = 10 \log A_2/A_1$$

where: NR = noise reduction (in decibels)
  A_1 = original room absorption (in sabins)
  A_2 = revised room absorption after material changes

Ducts and corridors Sound absorbing materials are primarily intended to control noise within the room itself, not to control transmission between rooms. However, absorption is valuable in reducing transmission through air ducts and corridors. In these cases, extensive use of absorbing materials such as acoustic tile or sound-absorbing liners may be required to properly dampen transmitted noise.

Estimating reverberation time

The addition of sound absorbing fabrics, upholstery, and other soft, porous materials will, of course, adversely affect the persistence of reflected sounds that are intended to reinforce the primary signal in lecture rooms, auditoriums, concert halls, and similar spaces that involve communication over significant distances.

As discussed in Chapter 1–2, this problem of sound persistence and blending is a basic consideration in the development of appropriate signal reinforcement. The objective is to produce full quality sound without excessive hardness or muddling. (See Tables 1–2.3, 1–2.4)

In design, *reverberation time* is manipulated by altering (1) room volume; (2) the absorbing or reflecting characteristics of surfaces and finishes; and (3) the density of room occupancy. This relationship is summarized in the following formula:

$$T_r = \frac{0.049 \ V}{\Sigma SA}$$

where: T_r = reverberation time (in seconds)
  (reverberation time is the interval required for a sound to decrease 60 decibels, or to one-millionth of its original intensity, after the source has stopped generating.)
  V = room volume (in cubic feet)
  ΣSA = room absorption, including occupancy (in sabins)

In critical situations, reverberation time should be calculated for each significant frequency. However, for preliminary estimates, a single calculation for 500 cps may be sufficient.

Echo control

Sound travels through room temperature air at a rate of approximately 1140 ft/sec. This relatively slow speed means that there may be a significant time lag between the generation of a sound and reception at the ear of a distant listener. (If the distance is 100 ft, the time lag will be approximately 1/11 or 0.09 sec.)

When the ear receives only directly transmitted sound signals, the integrity of the signals are unaltered by any time lag. But if the direct sound is implemented by reflections from room surfaces, the various components may arrive at the ear at different times. If the time differential is short (approximately 0.035 sec or less for speech), the direct and reflected components reinforce each other. But a longer time lag may produce muddling interference that will reduce intelligibility.

Distinctly audible echoes will occur when a listener hears a sufficiently

FIGURE 2–2.10

INFLUENCE OF REAR WALL

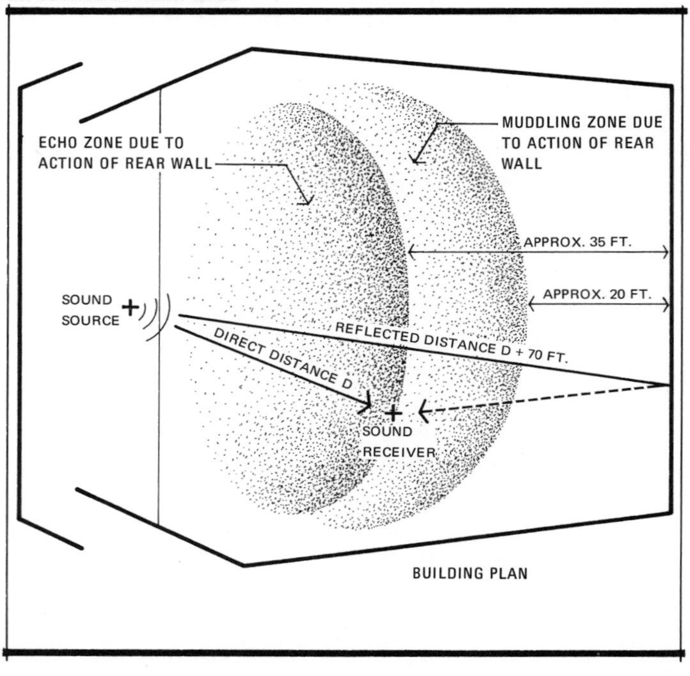

ECHO ZONE DUE TO ACTION OF REAR WALL

MUDDLING ZONE DUE TO ACTION OF REAR WALL

APPROX. 35 FT.

APPROX. 20 FT.

SOUND SOURCE

REFLECTED DISTANCE D + 70 FT.

DIRECT DISTANCE D

SOUND RECEIVER

BUILDING PLAN

intense reflected signal 0.06 sec or more after he hears the direct signal. This condition is most prevalent in large spaces where the reflected sound path is 65–70 ft (or more) longer than the direct path (see Figure 2–2.10).

In auditorium-type spaces, then, rear walls, ceiling areas adjacent to the front proscenium, or other surfaces that are remote from part of the audience may require special treatment to minimize or eliminate the reflected sound that produces the echo. This treatment will involve altering either the direction or intensity of the reflected sound. In this sense, commonly used techniques are: (1) minimizing the area of the offending surface, and making the surface that remains highly absorbent; (2) developing the offending wall as a diffusing surface, or tilting the surface to *ground* the reflection nearby; or (3) a combination of both surface modulation and absorption.

Flutter control There is a similar affect in simple rectangular spaces (such as offices or gymnasium areas) where the walls are reflective and unbroken. In these spaces, multiple horizontal reflections may occur between a single pair of parallel walls. A sharp impact, such as hand-clapping, will produce a ringing or buzzing sound that is called a *flutter echo*. This condition is most pronounced when the ear is at the same horizontal level as the sound source.

Flutter can be corrected by introducing draperies or similar absorbing materials on one of each pair of parallel walls. Alternative methods involve faceted wall panels or the use of diffusing objects such as large framed pictures, sculptural decoration, venetian blinds, etc.

CONTROL OF EXTERNALLY-GENERATED NOISE

The signal to background ratio is also affected by noise that originates in adjacent or remote spaces. These may be airborne sounds that induce subtle vibrations in the separating wall, ceiling, or floor. Or, if the sound source is in direct contact with the physical barrier, the vibrations are induced through direct impact.

Control involves: (1) the development of methods to reduce the intensity of induced vibrations, and (2) the development of barriers that will effectively resist transmission of such energy.

Material Action

The installation of absorbing materials on the ceiling or walls of the receiving space will have only a marginal effect in limiting the penetration of external noise. In fact, the porous and lightweight character of most sound absorbing materials seriously limits their value as sound insulators. Heavy materials are required, for the effectiveness of a sound isolating barrier will depend on: (1) its weight or mass, (2) its stiffness, and (3) its air tightness.

Weight, mass, and stiffness

Heavy partitions provide higher inertia to resist sound vibrations. As a general rule, then, the greater the mass, the greater the reduction in sound transmission.

In general, the transmission loss characteristics of a partition will improve at higher frequencies. However, isolation is also a function of stiffness, for a very stiff partition will not resist transmission as well as might be expected on the basis of weight alone. In practice, large panels often exhibit a dip in transmission resistance over a limited range of frequencies. The precise location of this *dip* will depend on the size of the panel and its bending stiffness. For many partitions that utilize stiff, homogeneous panels, however, this *resonance dip* can be as much as 10–15 db in intensity and will occur within the speech range; thus permitting a higher than expected transmission of speech signals.

Air space separations

An isolating air space that separates individual partition layers will help to reduce the transmission of vibrations from one side of a barrier to the other. This method makes it possible to use lightweight construction methods while maintaining good resistance to sound transmission. It is also effective for reducing the previously discussed resonance dips.

In both cases, care must be taken to insure that structural contact across the air space is minimized (structural discontinuity).

Transmission loss

Wall and ceiling-floor assemblies are rated according to their ability to resist the transmission of airborne sounds. The more general method of rating is the *Sound Transmission Class* (STC), which utilizes the criteria curves discussed in Chapter 1–2 (see Figure 1–2.10).

Figure 2–2.11 summarizes the performance of several typical assemblies. When interpreting these, note that normal speech can usually be heard clearly through a partition that is rated near STC 35. The same

FIGURE 2–2.11

BASIC CONSTRUCTION PRINCIPLES

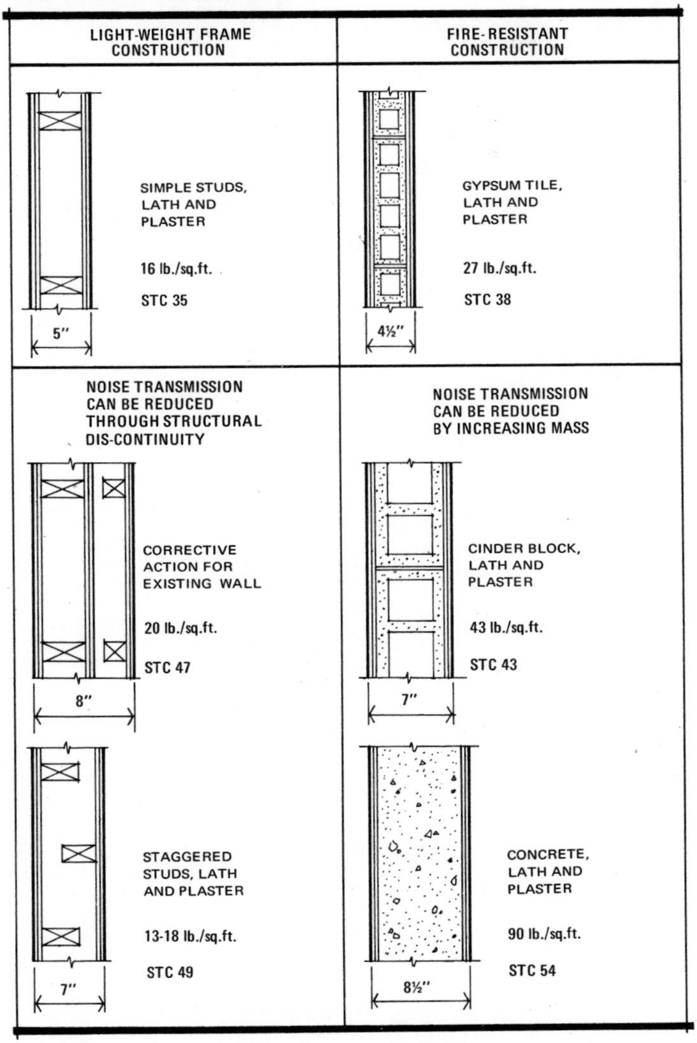

LIGHT-WEIGHT FRAME CONSTRUCTION	FIRE-RESISTANT CONSTRUCTION
SIMPLE STUDS, LATH AND PLASTER 16 lb./sq.ft. STC 35 5"	GYPSUM TILE, LATH AND PLASTER 27 lb./sq.ft. STC 38 4½"
NOISE TRANSMISSION CAN BE REDUCED THROUGH STRUCTURAL DIS-CONTINUITY CORRECTIVE ACTION FOR EXISTING WALL 20 lb./sq.ft. STC 47 8"	NOISE TRANSMISSION CAN BE REDUCED BY INCREASING MASS CINDER BLOCK, LATH AND PLASTER 43 lb./sq.ft. STC 43 7"
STAGGERED STUDS, LATH AND PLASTER 13-18 lb./sq.ft. STC 49 7"	CONCRETE, LATH AND PLASTER 90 lb./sq.ft. STC 54 8½"

speech will be heard only as a murmur through an STC 40 wall, and will be negligible through an STC 50 wall. (Also see Table 1–2.7.)

When more precise and specific analysis is required, transmission loss data is generally available for common construction details (see Table 2–2.4).

Table 2-2.4 Transmission Loss Values

Representative Construction Details	TRANSMISSION LOSS (db)					
	125 cps	250 cps	500 cps	1000 cps	2000 cps	4000 cps
Single wall: 2 × 4 wood studs, ½-in plaster on ⅜-in gypsum lath, both sides (16 lb/sq ft)	27	25	31	44	34	50
Single wall: 6-in hollow-core cinder block, ⅝-in plaster, both sides (43 lb/sq ft)	36	33	38	45	50	56
Single wall: 7-in concrete, plastered both sides (90 lb/sq ft)	44	42	52	58	66	70
Double wall: Staggered 3¼-in wire studs, ½-in plaster on ⅜-in gypsum lath, both sides (5¾-in total thickness; 13 lb/sq ft)	34	39	44	49	48	60
Double wall: 4-in hollow-core gypsum block, ⅝-in plaster on one side; ½-in plaster on ⅜-in gypsum lath on second side; ⅞-in resilient metal clips connect two sides; (26 lb/sq ft)	25	33	43	48	49	54
Double wall: Two sections of plastered 4½-in solid brick; 2-in air space separation; sound absorbing material in air space; bridging at edges only (90 lb/sq ft)	43	50	52	61	73	78
Floor-ceiling: Typical residential section; subfloor and finish floor on wood joists; gypsum lath and plaster below (c. 15 lb/sq ft)	24	32	40	48	51	54
Floor-ceiling: 3½-in concrete slab, ½-in plaster below (c. 45 lb/sq ft)	43	40	44	53	56	58
Door: 1⅜-in hollow-core door, normally hung	5	11	13	13	13	12
Door: 1⅜-in solid wood door, normally hung	10	13	17	18	17	15
Door: 1⅜-in solid wood door, fully gasketed	16	18	21	20	24	26

Estimating the effect of composite barriers

Most walls include more than one type of construction. This difference may involve a door or window section that has a noise reduction value (TL_o) below the value assigned to the greater portion of the barrier (TL_w). (If the minor area is a hole or opening, $TL_o = 0$.)

The effective or average transmission loss of the composite wall (TL_e) can be estimated from Figure 2–2.12.

FIGURE 2–2.12

EFFECTIVE TRANSMISSION LOSS OF A COMPOSITE BARRIER

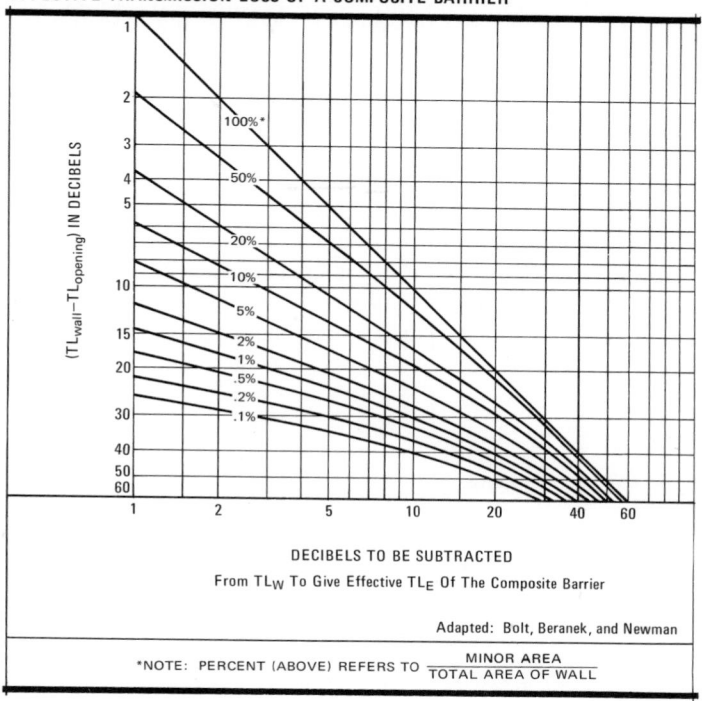

DECIBELS TO BE SUBTRACTED

From TL_W To Give Effective TL_E Of The Composite Barrier

Adapted: Bolt, Beranek, and Newman

*NOTE: PERCENT (ABOVE) REFERS TO $\dfrac{\text{MINOR AREA}}{\text{TOTAL AREA OF WALL}}$

Estimating noise control value (external noise)

The effective transmission loss of the barrier is the most significant single factor in determining the intensity of transmitted sounds. To a lesser extent, however, intensity is also affected by the area of partition surface, and by the room absorption of the receiving room. In this sense, the precise spatial effect in the receiving room can be estimated with the following formula:

$$NR = TL_e - 10 \log S/A_2$$

where: NR = noise reduction (in decibels)

   TL_e = effective transmission loss of the homogeneous or composite wall (in decibels)

   S = the area of common surface between the *source room* and the *receiving room* (in square feet)

   A_2 = room absorption of the *receiving room* (in sabins)

Note that if the room absorption (A_2) is low relative to the surface area (S), the effect of the receiving room is to diminish the transmission loss value (TL_e). However, if the room absorption numerically exceeds the surface area, the effect of the receiving room is to increase the total noise reduction.

Flanking noise

No matter how good an insulator a partition may be, cracks and holes can very quickly nullify its value. Obviously the design and placement of doors and windows are major considerations. But poorly designed ceiling cavities, wall-to-ceiling or wall-to-floor cracks, back-to-back switch boxes, and similar paths can adversely affect the insulation value of a partition (see Figure 2–2.13). These paths must be designed to minimize sound transmission and leakage.

FIGURE 2–2.13

EFFECT OF CRACK OPENINGS

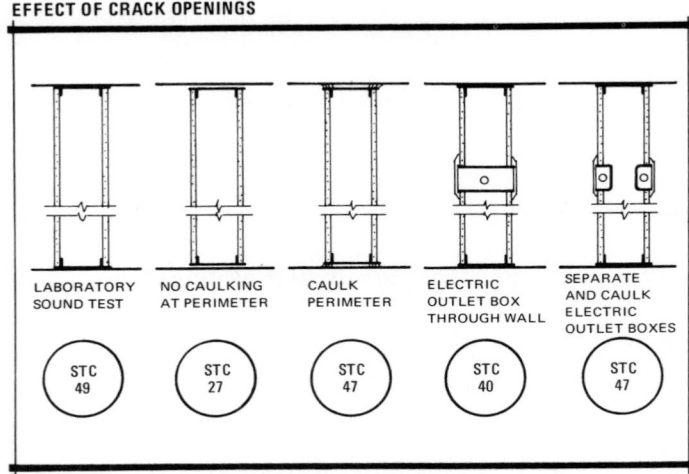

LABORATORY SOUND TEST	NO CAULKING AT PERIMETER	CAULK PERIMETER	ELECTRIC OUTLET BOX THROUGH WALL	SEPARATE AND CAULK ELECTRIC OUTLET BOXES
STC 49	STC 27	STC 47	STC 40	STC 47

FIGURE 2–2.14

METHODS OF CAVITY CLOSURE

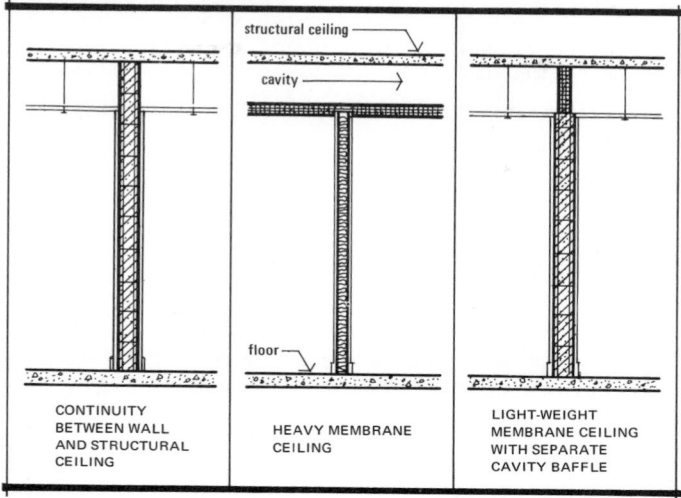

CONTINUITY
BETWEEN WALL
AND STRUCTURAL
CEILING

HEAVY MEMBRANE
CEILING

LIGHT-WEIGHT
MEMBRANE CEILING
WITH SEPARATE
CAVITY BAFFLE

FIGURE 2–2.15

REDUCTION OF IMPACT TRANSMISSION

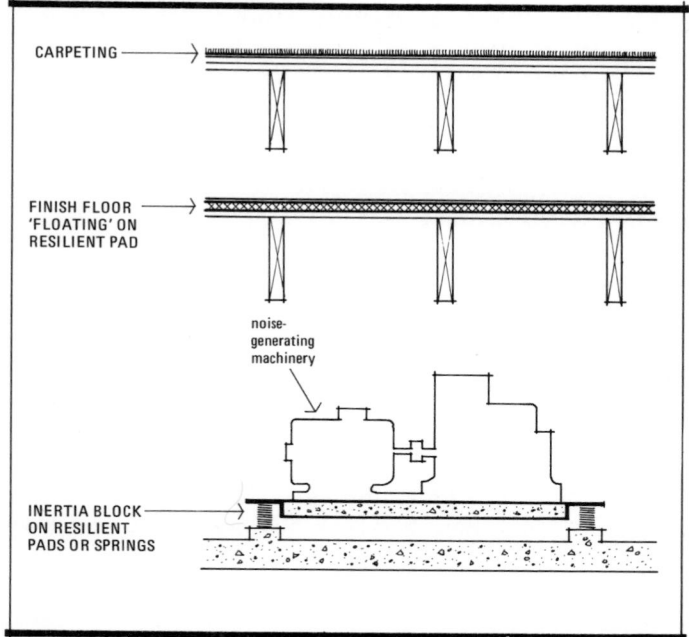

CARPETING

FINISH FLOOR
'FLOATING' ON
RESILIENT PAD

noise-
generating
machinery

INERTIA BLOCK
ON RESILIENT
PADS OR SPRINGS

Structural Impact

When an intermittent sound source comes into direct contact with the barrier (such as occurs with footsteps or simple hammer action), transmission is reduced by inserting a resilient material between the source and the barrier.

Structural transmission may also involve sustained vibrations (such as fan vibrations or vibrations caused by a rotating or reciprocating engine). When such vibrations are subtle, the insertion of resilient padding may again be sufficient for control. For more intense vibration, springs may be inserted between the source and the barrier (see Figure 2–2.15).

As a simplified procedure for preliminary estimating purposes in general-use spaces, floor assemblies are rated according to their ability to resist the transmission of more typical impact noises. This is the *Impact Noise Rating* (INR), which is discussed in Chapter 1–2.

Table 2-2.5 Impact Noise Rating

Representative Construction Details	*Typical STC*	*Typical INR*
Reinforced 7-in concrete slab; tile floor	49	−17
Same concrete slab; add suspended gypsum ceiling		− 4
Same concrete slab; add 2-in concrete floor floating on 1-in resilient pad		+ 1
Same concrete slab; add 20-oz wool carpet		+ 2
Same concrete slab; add wood floor floating on 1-in resilient pad		+ 7
Steel bar joists; 2-in concrete slab floor; plaster on gypsum lath ceiling	48	−10
Same joists, slab, and ceiling; add carpet on foam rubber pad		+20

Table 2–2.5 summarizes the performance of several typical assemblies. When interpreting these, note that a +INR indicates a tendency toward superior performance. A −INR indicates a tendency toward inferior performance.

MECHANICAL NOISE

Noises that are associated with mechanical equipment can be classified in several categories:

1. *airborne equipment noise that is transmitted to occupied spaces through common walls or ceiling-floor assemblies*

This is primarily controlled through partition or barrier design, to insure sufficient mass, minimum stiffness, and possibly structural discontinuity (air space separations). All joints and cracks must be effectively sealed.

2. *airborne equipment noise that travels through air supply and return ducts or plenums*

Generally, this is controlled by placing absorbing blankets and linings within the duct or plenum. However, particularly for short duct lengths and wall openings, special silencers or mufflers may be required (see Figure 2–2.16).

FIGURE 2–2.16

SOUND 'TRAPS'

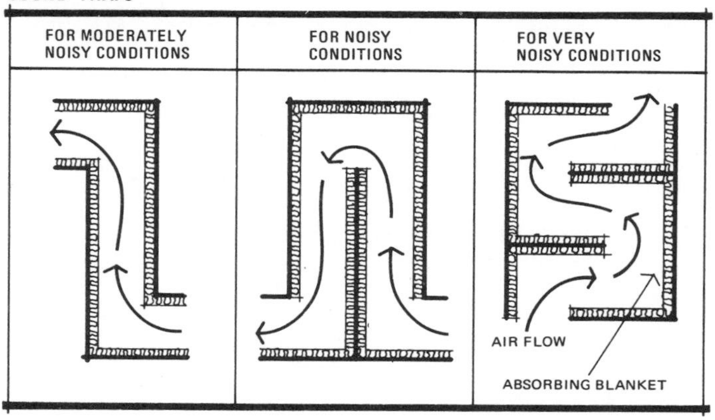

| FOR MODERATELY NOISY CONDITIONS | FOR NOISY CONDITIONS | FOR VERY NOISY CONDITIONS |

AIR FLOW

ABSORBING BLANKET

3. *structural-borne sounds induced by vibrating equipment that is rigidly attached to the structure or barrier*

This is generally controlled through the use of rubber mounting pads or steel springs. Note that it may be necessary for these devices to absorb deflections of up to 6 in. for powerful low-rpm equipment.

A related method of treatment involves hanging the equipment from the ceiling with vibration hangers.

Care must be taken to isolate all paths to and from the equipment, including the provision of resilient or flexible couplings for pipes, ducts, and conduit.

4. *internal noise induced by air supply diffusers and grilles that are located within the room itself*

The rated noise level of the mechanical device itself is usually the responsibility of the manufacturer. So this aspect is primarily controlled by preparing a suitably rigid performance specification to guide equipment selection (see Chapter 1–2 for a discussion of Noise Criteria Curves).

When suitably-rated devices are not available, air noise can be further controlled by reducing large pressure drops within the distribution system. Noise reductions can be realized, then, by selecting an oversized unit, with the intention of operating it below normal rated capacity in order to reduce the sound level.

A second alternative is to select several smaller units. When this is done, however, care should be taken to insure that the cumulative noise effect does not exceed the applicable Noise Criteria Curve. (See Figure 2–2.7 for estimating the additive effect of multiple noise sources.)

Calculated and Graphic Estimates of Acoustical Performance

When estimating the comprehensive acoustical performance of a space, Table 2–2.6 is useful for summary purposes.

Table 2-2.6 Estimation of Acoustical Performance (Intensity)

	Formulas	References
Noise Reduction:		
Room Absorption	$\Sigma SA = S_1A_1 + S_2A_2 + S_3A_3 + \dots$	See Table 2–2.3 See Figures 1–2.4 and 1–2.8 for criteria
Wall and Ceiling Barriers (transmission)	Obtain curve of transmission loss (or STC rating)	See Table 2–2.4 See Figure 1–2.10 and Table 1–2.7 for criteria
Floor-Ceiling Barriers (impact)	Obtain curve of impact noise transmission (or INR rating)	See Table 2–2.5 See Figure 1–2.11 and Table 1–2.8 for criteria
Internal Operating Equipment	Obtain curve of operating equipment noise (or NC rating)	See Figure 1–2.9 and Table 1–2.6 for criteria
Signal Reinforcement:		
Natural Amplification	Develop ray diagrams	See Figure 2–2.6 and Table 2–2.1
Control of Echoes and Muddling	Analyse ray diagrams	See Figure 2–2.10
Reverberation Time	$T_R = \dfrac{0.049\ V}{\Sigma SA}$	See Tables 1–2.3 and 1–2.4 for criteria (also Figure 1–2.6)

Thermal and Atmospheric Control

Heat energy is generated by the random motion of molecules, and the intensity of this molecular action is measured by thermometer scales. A temperature of *absolute zero* ($-459°F$ or $-273°C$) is the point at which heat is no longer being generated because all molecular motion stops. The unit measure ($1°F$; $1°C$) relates to the quantity of heat energy required to raise a given quantity of water one step on the scale (one BTU raises one pound of water from $60°F$ to $61°F$; one calorie raises one gram of water from $15°C$ to $16°C$).

As the intensity of molecular action increases, the molecular structure becomes more diffused, and the matter will assume different physical forms. Generally, this is a progressive change from a *solid* to a *liquid* (water changes at $32°F$) to a *gas* (water changes at $212°F$).

Heat is analogous with liquids and electricity in the sense that a difference in intensity (or pressure) produces the potential for flow. Heat always flows from a warm object or mass to a cooler one (i.e., from the higher rate of molecular action to the lower rate). This flow of energy continues until a neutral or equal temperature condition exists between the two. A lighted lamp, for example, is warmer than the surrounding air and warmer than the body surface of a human being. Since thermal gradients exist, a flow of heat results. Under extreme conditions, this heat can produce discomfort for the occupant.

Heat transfer related to human comfort takes place in several ways:

(1) *Radiant heat* is short wave infrared energy that passes as a ray or beam through air with little absorption or dispersion. When energy in this electromagnetic form strikes a person, a surface, or an object, heat is absorbed or reflected (in a manner similar to the action of light), and a rise in surface temperature results due to the absorbed energy. This form of heat is difficult to control with conventional air circulation methods because the transfer of heat is directly between

two objects or surfaces, and the surrounding air exerts little influence in this process.

Control of heat in this form, then, depends on techniques that manipulate the temperature gradients between surfaces (or between surfaces and energy sources). This involves: (1) techniques that adjust the overall relationship of surface temperatures within an enclosed environment, relating the temperature of these surfaces to the surface temperature of the human skin; or (2) techniques that modify the directional concentration of radiant energy, such as source reflectors or shading devices.

(2) The second type of heat transfer *does* respond to conventional air circulation methods and is the most prominently used method of environmental control for either heating or cooling. Heat is transferred directly from a hot object to the surrounding air (*conduction*); the warmed air then moves by gravity *convection* (hot air rises) or by forced circulation to a cooler object, where heat is again transferred by conduction. This last transfer warms objects and surfaces within the room.

The reverse conditions are applied when environmental cooling is required.

The general characteristics of convective motion are involved whenever heat is borne by a liquid (such as hot water), by environmental air (such as warm or cool air), or by a gas (such as steam). The rate of conductive heating or cooling depends on the velocity of the circulating medium, and on the difference in temperature between the medium and the surface to be heated or cooled.

(3) Although thermal adjustment of the human body is accomplished primarily by convection and radiation when environmental temperatures are low or moderate, *evaporation* becomes an increasingly important controlling factor when environmental temperatures are high. In fact, at high temperatures, air and most untreated room surfaces tend to lose their cooling potential, and the human body will depend almost completely on evaporation in its effort to maintain thermal equilibrium.

Evaporation of surface moisture (such as perspiration) will cool any object or body because heat is taken from the mass to facilitate the evaporative process.

For this reason, control of water vapor in the environmental air is an important qualitative aspect. Humidity should be low enough in warm weather to facilitate evaporative cooling of the skin, and high enough in dry weather to prevent excessive drying or dehydration.

These three forms of heat transfer affect and are the means for manipulating the thermal environment for human occupancy. This habitable environment is partially achieved through the development of an enclosing building shell; and it is further balanced with heat-generating and heat-consuming devices. Both of these concepts must be evaluated in terms of their response to the local climatic environment.

THE SOLAR ENVIRONMENT

The earth revolves about the sun in an approximately circular path, with the sun located slightly off-center in this circle. The earth is nearest to the sun about January first, and is at the more remote position about July first.

As a result of this slightly eccentric position of the sun, the intensity of solar radiation at the outer limit of the earth's atmosphere is about 7% higher on January first (approximately 445 btu/sq ft/hr) than it is on July first (approximately 415 btu/sq ft/hr). This represents a fluctuation of 3.5% from the *mean* solar radiation at the outer limit (generally taken to be 420–430 btu/sq ft/hr).

As this radiation passes through the atmosphere, part of the energy is reflected and scattered by dust particles and by water and air molecules. This scattered energy becomes the diffuse sky vault.

In addition to this scattering effect, part of the solar radiation is absorbed by ozone in the upper atmosphere and by water vapor near the surface.

Atmospheric variables

The energy that is incident at the surface of the earth is therefore present in two forms: (1) direct radiation that has been somewhat diminished in its passage through the atmosphere, and (2) diffuse sky radiation.

In this respect, as the atmosphere becomes congested with clouds, dust, or smoke and other industrial contaminants, the direct radiation is increasingly diminished. For example, on a clear day at sea level, the proportion of direct radiation (directional sunlight) to diffuse radiation (nondirectional skylight) is typically 70:30. In many large cities, where the atmosphere is contaminated by industrial gases and particles, this ratio will change. In this situation, a clear day ratio of 45:55 or less is more typical.

On overcast days, the diffuse component will increase further, with a corresponding decline in direct radiation; until, with heavy cloudiness, nearly all of the incident energy is in the diffuse (nondirectional) form.

But in addition to the diffusing effects of clouds and contaminants, intensity changes are also inherent with clear sky conditions because direct solar energy will vary with the depth of the air mass through which the rays must pass. This depth of passage is thinnest at noon in summer, when the sun is nearly overhead. In this situation, an approximate 294 btu/sq ft/hr is incident on a surface normal to the plane of the sun. At 3 PM on the same day, the slightly more oblique passage of energy (caused by rotation of the earth relative to the sun) reduces the maximum at the surface to 266 btu/sq ft/hr. At 6 PM, passage through the atmosphere is much more oblique, and the maximum penetration is reduced to 67 btu/sq ft/hr.

Geographic and Climatic Variations

If the earth is evaluated as a thermodynamic system, with outer space as the external environment, radiation is the only form of heat transfer that can maintain an overall heat balance. In this regard, it has been estimated that as a general average over the entire earth, 43% of the solar energy that is incident at the outer limit of the atmosphere reaches the earth's surface. After it absorbs radiant energy, then, the earth itself becomes a secondary radiant source, emitting energy toward outer space.

Of course, solar gains and losses will differ considerably with local conditions. The most significant of these conditions relates to the variable and nonuniform distribution of solar energy that is caused by the 23.5° tilting of the axis of the earth. This produces *seasonal* variations in the altitude of the sun above any specific location on earth. During winter periods, it also exaggerates the fact that the more northern (or southern) latitudes receive sun rays at an angle of incidence that is much more oblique than it is near the equator. (As noted previously, oblique passages through the atmosphere produce significant reductions in the penetration of energy.)

Because of these effects, some researchers have concluded that the portion of the earth's surface that is defined by 30°N and 30°S exhibits a net gain in heat by radiation, while the two segments between 30° latitudes and the poles exhibit a net loss by radiation.

Without atmospheric circulation, therefore, the regions near the equator would become progressively hotter, while the areas of higher latitude (north and south) would become progressively cooler. The general wind patterns and ocean currents provide the primary means for transporting heat from the equator toward the poles.

Atmospheric Heat Storage

When the earth emits energy toward outer space, some of this energy penetrates the atmosphere and escapes. But the greater portion (approximately 70%) is intercepted by water vapor in the atmosphere and is returned again to earth by radiation or convection. On overcast days, the clouds form a local barrier that minimizes losses from the earth's surface. Thus, the water vapor in the atmosphere exerts a *greenhouse effect,* retaining heat and stabilizing temperatures at the surface of the earth.

This stabilizing influence is somewhat seasonal in that it causes peak temperature conditions to lag behind peak solar conditions. For example, the heat storage effects that produce maximum summer temperatures in most of the United States do not generally prevail until several weeks after the maximum solar gains (which occur about June 21, when the sun reaches its most northerly position). Sustained peak temperature conditions are generally judged to occur in early August.

Sustained minimum temperature conditions are similarly judged to occur in late January or February—several weeks after the minimum solar energy condition (which occurs at about December 22, when the sun reaches its most southerly position).

From these four approximate dates, the conditions of maximum heat gain and maximum heat loss can generally be derived for a given location. These maximum conditions depend: (1) on heat transfers that are associated with solar radiation, and (2) on heat transfers that are associated with the atmospheric air mass.

MAN-MADE ENCLOSURES

Buildings are, in part, a response to the variable solar and climatic influences on earth. In this sense, the aspect of thermal comfort is a significant criteria in the development of the building enclosure. The term *comfort* can here be broadly defined as the condition where thermal stress is minimized; where the occupant can adjust to his environment with a minimum expenditure of automatic body effort. In this sense, the thermal function of the enclosure is to assist in the establishment of an area of biological equilibrium.

In its basic form, the enclosure functions to admit and conserve heat when the external environment is cold; and to impede or dissipate the penetration when the external conditions are warm.

This thermal function is most readily observed in the more extreme climatic regions. For example, the evolution of the domed *igloo* by the Eskimos was their response to prolonged and extreme cold weather conditions in areas where materials are scarce. This snow-packed hemispherical form is compact (i.e., it is a minimum-area enclosure surface); it is well-sealed and insulated by the snow; and the form effectively deflects winds, with entrance openings oriented away from the prevailing winds to reduce drafts.

Many subarctic structures are somewhat similar in concept in that they have common *party walls* and minimum-surface exterior shapes. These characteristics combine to reduce the exposed surface area of each habitable unit. There are also variations of double-shell construction techniques, which provide an effective insulating barrier with minimum utilization of materials.

At the other extreme, it is interesting to note that hot and arid regions (i.e., semidesert) have tended to evolve somewhat similar enclosure concepts. For example, the Pueblo Indians developed large communal structures that utilize *party walls* to reduce exposed surface area. They were not overly concerned about insulation, but massive adobe walls acted to delay heat transmission; while small window and door openings were placed to minimize solar exposure.

Domed roof structures are also common in these hot, arid regions. This prevalence is a response to the intense and unshaded solar loads that are incident on roof areas in semidesert areas. When a dome is substituted for a flat roof, the roof area becomes several times the base area, which means that average solar radiation is somewhat reduced by the larger curved surface. The thermal action is further implemented by winds that tend to dissipate heat from the enlarged surface; and where heavy masonry materials store heat by day, night breezes will help to dissipate that heat before it penetrates to the interior.

Each of these regional building forms evolved from the need to control (1) convection-conduction transfers through the enclosure, and (2) radiant transfers to and from the occupant caused by cold or warm peripheral enclosure surfaces. However, building in hot and humid regions (such as the tropics) are faced with a very different problem. In these regions, evaporation of moisture for dampness control and for occupant cooling effect is very important. In tropical areas, habitable structures have minimum walls to make maximum use of penetrating breezes for evaporation control. Furthermore, floors may be elevated to facilitate ventilation. Tree cover and roof structures are then developed to provide shade and insulation (and roof overhangs are developed to protect the minimum walls against rain).

It is in temperate regions that the greatest potential for structural diversity exists. Since these zones produce extremely variable climatic conditions, the building may periodically be called upon to respond in a manner similar to each of the preceding regional categories; at times providing an effective barrier to exclude extreme cold or extreme heat, at other times providing shade plus openness to permit the penetration of cooling breezes. This conflict of criteria to meet temporary and seasonal changes in the basic external environment has produced the considerable diversity of form that is found in these temperate regions. It has also produced the stimulus to develop mechanical devices to compensate for the compromises that must inevitably be made in regard to form, materials, and orientation of openings.

This requires that builders in temperate regions must distinguish between periods when man seeks solar influences (i.e., cold weather) and periods when he seeks to avoid them; between periods when winds are beneficial and periods when winds should be intercepted. Building orientation is often a compromise, and materials, forms, and mechanical devices should be used accordingly.

Heat Transfer Through the Building Shell

A number of factors influence an attempt to evolve satisfactory enclosures that respond to the varying solar and climatic conditions in the north temperate zone. Basically, these fall into the following general categories:
(1) instantaneous heat gains associated with solar energy that passes through transparent openings in an enclosure shell
(2) delayed heat gains associated with solar energy that impinges on opaque portions of the shell
(3) heat gains or losses associated with conductive transfer through the shell, to or from adjacent outdoor air
(4) heat gains or losses associated with unintentional infiltration of air through cracks and openings
(5) heat gains or losses associated with intentional expulsion of stagnant or contaminated indoor air, and replacement with fresh outdoor air.

Penetration of solar radiation through openings

The transmission of solar energy through openings in the enclosing shell is nearly instantaneous. The intensity of this transmission at a given point and time will depend: (1) on the orientation of the opening, and (2) on the reflection and absorption characteristics of the transmitting material (if any).

For roof openings, such as a skylight, the atmospheric air mass is thinnest (and therefore offers the least resistance) when the sun is directly overhead. This is the time when rays strike the horizontal roof at something approaching normal incidence; so direct solar radiation through these openings approaches a daily maximum at solar noon.

Vertical wall openings rarely receive direct radiation at near-normal incidence. The more oblique passage of rays through the atmospheric air mass tends to reduce intensity somewhat. This means that radiation transmitted through windows and doors is generally somewhat lower in maximum intensity than that transmitted through roof skylights. (See Tables 2–3.1, 2–3.2, and 2–3.3.)

Table 2-3.1 Typical Solar Design Data—1

December 22—(40° North Latitude)

	Solar Time	Location of Sun Azimuth[1]	Location of Sun Altitude
Sunrise	7:30	121°0'	0°0'
	8:00	127°0'	5°30'
	Noon	180°0'	26°30'
	4:00	127°0'	5°30'
Sunset	4:30	121°0'	0°0'

[1]*Azimuth* refers to the horizontal angle from north.

	Approx. Solar Heat Gain through Clear, Unshaded Glass (btu/sq ft/hr)		
	8 AM	Noon	4 PM
Typical overcast condition	2	15	2
Typical clear day condition			
N	2	15	2
NE	6	15	2
E	60	16	2
SE	75	159	3
S	44	230	44
SW	3	159	75
W	2	16	60
NW	2	15	6
Roof	7	102	7

The more significant comfort control problems created by enclosure openings occur when intensity of solar radiation is near maximum. At these times, direct rays must generally be tempered in the immediate vicinity of the human occupant. This can be accomplished in two ways: (1) by orientation of openings to minimize the penetration of direct solar radiation, or (2) by shading those openings that for other reasons must remain exposed to direct radiation at a time when this may become an adverse influence on comfort or on basic system performance.

Orientation of walls and openings to respond to solar load For many combinations of orientation and latitude, a wall or opening will re-

Table 2-3.2 Typical Solar Design Data—2

March 21, September 21—(40° North Latitude)			
	Solar	*Location of Sun*	
	Time	*Azimuth*[1]	*Altitude*
Sunrise	6:00	90°0'	0°0'
	8:00	110°30'	22°30'
	Noon	180°0'	50°0'
	4:00	110°30'	22°30'
Sunset	6:00	90°0'	0°0'

[1]*Azimuth* refers to the horizontal angle from north.

	Approx. Solar Heat Gain through Clear, Unshaded Glass (btu/sq ft/hr)		
	8 AM	*Noon*	*4 PM*
Typical overcast condition	14	25	14
Typical clear day condition			
N	14	25	14
NE	81	25	14
E	185	27	14
SE	179	131	14
S	64	185	64
SW	14	131	179
W	14	27	185
NW	14	25	81
Roof	76	201	76

ceive only grazing radiation. Others will receive radiation at more nearly perpendicular angles, and these are the more critical exposure surfaces.

In the north temperate zone, heavy radiation loads will act most decisively on the roof and on the east and west exposures during the summer period. South exposures permit moderately significant heat gains during the summer; but they permit very significant heat gains during the winter. North exposures receive minimal radiation throughout the year. (See Figures 2–3.2 and 2–3.3.)

To be somewhat more specific:

(1) If they receive direct radiation at all, walls facing north of east or north of west will tend to receive this radiation only in the late spring and early summer.

(2) Walls facing south of east or south of west will tend to receive maximum direct radiation in the late fall and early winter.

Table 2-3.3 Typical Solar Design Data—3

June 21—(40° North Latitude)

	Solar Time	Location of Sun	
		Azimuth[1]	Altitude
Sunrise	4:30	59°0'	0°0'
	8:00	89°0'	37°30'
	Noon	180°0'	73°30'
	4:00	89°0'	37°30'
Sunset	7:30	59°0'	0°0'

[1]*Azimuth* refers to the horizontal angle from north.

	Approx. Solar Heat Gain through Clear, Unshaded Glass (btu/sq ft/hr)		
	8 AM	Noon	4 PM
Typical overcast condition	23	34	23
Typical clear day condition			
N	26	34	26
NE	140	34	23
E	194	37	23
SE	137	64	23
S	26	85	26
SW	23	64	137
W	23	37	194
NW	23	34	140
Roof	138	240	138

Table 2-3.4 Solar Time Variations

Representative Geographical Approximations which Define the 40°N Latitude Line (Used as a typical condition in this text)			Approximate Time of Solar Noon
Philadelphia, Pa.	40°N	75°W	12:00 EST
Columbus, Ohio	40°N	83°W	12:30 EST
Springfield, Ill.	40°N	90°W	12:00 CST
			(1:00 EST)
Manhattan, Kans.	39°N	97°W	12:30 CST
Boulder, Colo.	40°N	105°W	12:00 MST
			(2:00 EST)
Salt Lake City, Utah	41°N	112°W	12:30 MST
Reno, Nev.	40°N	120°W	12:00 PST
			(3:00 EST)

FIGURE 2–3.1

AZIMUTH AND ALTITUDE ANGLES

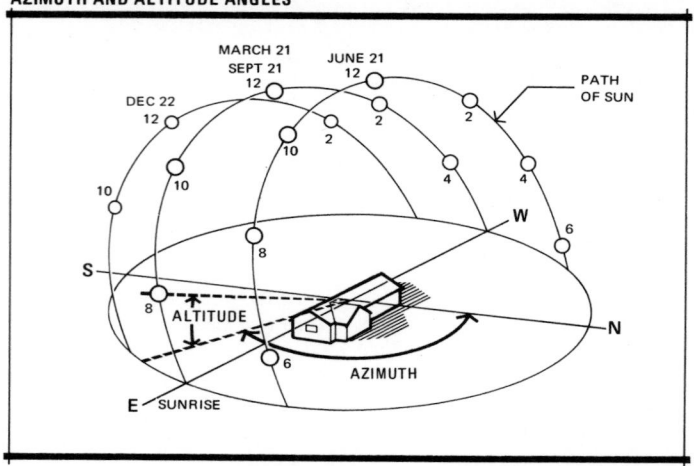

FIGURE 2–3.2

EFFECT OF SOLAR HEAT GAIN THROUGH GLASSED OPENINGS:
PEAK WINTER PERIODS

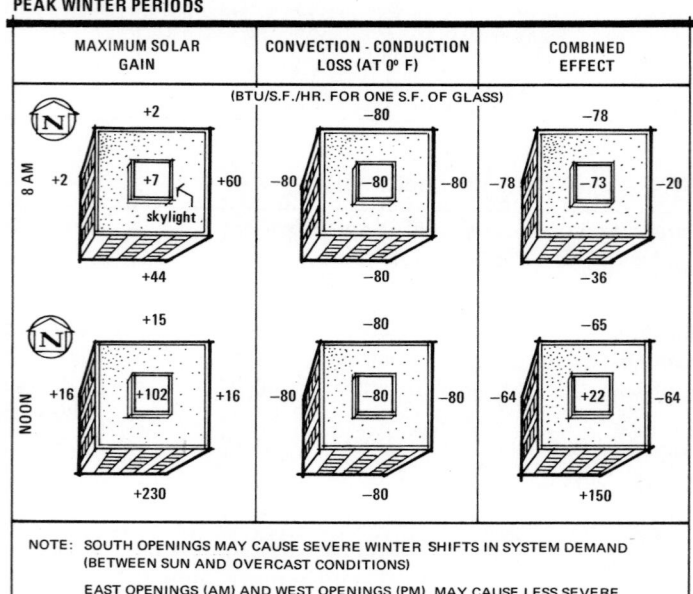

NOTE: SOUTH OPENINGS MAY CAUSE SEVERE WINTER SHIFTS IN SYSTEM DEMAND
(BETWEEN SUN AND OVERCAST CONDITIONS)

EAST OPENINGS (AM) AND WEST OPENINGS (PM) MAY CAUSE LESS SEVERE
WINTER SHIFTS

FIGURE 2–3.3

EFFECT OF SOLAR HEAT GAIN THROUGH GLASSED OPENINGS:
PEAK SUMMER PERIODS

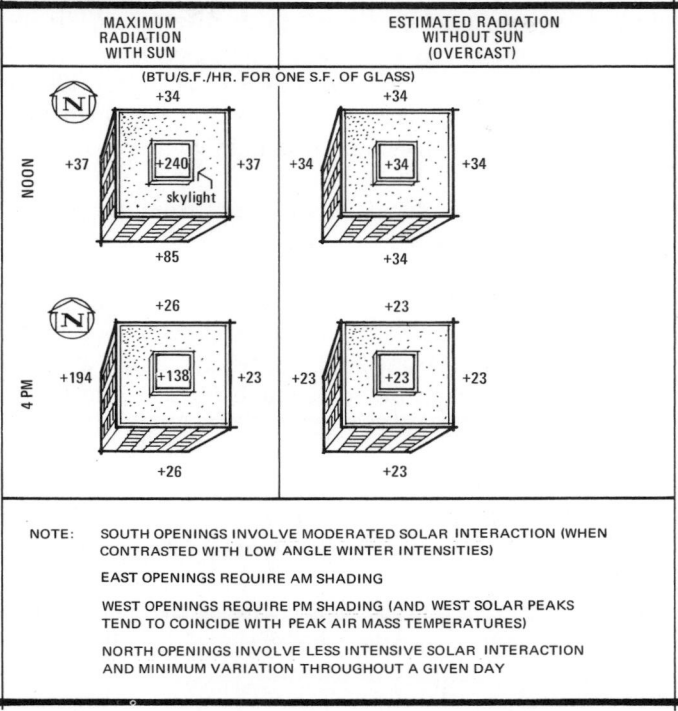

MAXIMUM RADIATION WITH SUN	ESTIMATED RADIATION WITHOUT SUN (OVERCAST)

(BTU/S.F./HR. FOR ONE S.F. OF GLASS)

NOON

+34
+37 +240 skylight +37
+85

+34
+34 +34 +34
+34

4 PM

+26
+194 +138 +23
+26

+23
+23 +23 +23
+23

NOTE: SOUTH OPENINGS INVOLVE MODERATED SOLAR INTERACTION (WHEN CONTRASTED WITH LOW ANGLE WINTER INTENSITIES)

EAST OPENINGS REQUIRE AM SHADING

WEST OPENINGS REQUIRE PM SHADING (AND WEST SOLAR PEAKS TEND TO COINCIDE WITH PEAK AIR MASS TEMPERATURES)

NORTH OPENINGS INVOLVE LESS INTENSIVE SOLAR INTERACTION AND MINIMUM VARIATION THROUGHOUT A GIVEN DAY

(3) Walls facing east of north or east of south will tend to receive maximum direct radiation at sunrise or during the morning hours.

(4) Walls facing west of north or west of south will tend to receive maximum direct radiation during the afternoon hours or at sunset.

Openings in the east and west walls are therefore subject to direct radiation loads that vary considerably during the day. These openings permit particularly severe penetrations early and late in the day, because low sun angles create conditions in which there are increased effective window cross sections perpendicular to the rays of the sun. Furthermore, with west walls, peak solar gains occur coincidentally with high afternoon air temperature and humidity conditions during summer periods. This means that west openings admit heavy solar gains at the same time that peak ventilation, infiltration, and conduction heat gains occur.

As a result, the nature of these maximum radiation loads suggests that buildings in the north temperate regions are generally best developed so that most major openings are oriented toward the north and south, with reduced east or west exposure.

Shading techniques and devices   In relative terms, direct solar radiation constitutes one of the most significant thermal influences on the building shell. Because the character of this load changes with time of day and season of the year, perimeter locations within the enclosure will constantly undergo thermal changes, and occupant comfort may be adversely affected on a localized basis.

The principal thermal problem here derives from the fact that clear glass permits almost unimpeded penetration of radiant energy. To minimize extreme variations in thermal performance among walls of differing orientation, therefore, some wall assemblies may be modified to intercept direct radiation and minimize its penetration into the occupied interior.[1] This treatment can take several forms, and the basic alternatives that follow are listed in increasing order of reliability and effectiveness.

Interior Blinds and Shades.   While generally effective for shading room occupants from direct radiation, interior shading devices permit the radiant heat to enter the space itself before it is intercepted. As a result, the absorbed energy may become an adverse thermal influence through interior convection and reradiation.

However, light-colored finishes will increase the effectiveness of the shading system because a greater proportion of the incident energy is reflected back through the glass to exterior space.

Absorbing and Reflecting Glass.   Clear glass can be replaced with reflecting or absorbing glass. In general, reflecting glass types will be more effective than absorption types because more energy tends to be dissipated (by reflection) back toward the outdoor space; while absorbed energy is dissipated in both directions by convection and reradiation.

Adjacent Structures and Landscaping.   Appropriately placed natural and man-made elements can be utilized. Adjacent trees and buildings can

---

[1] The *shading coefficient* includes consideration of direct transmission, absorption, and reradiation, and is the ratio of:

$$\frac{\text{Total Solar Heat Gain With Shading}}{\text{Total Solar Heat Gain Without Shading}}$$

The *shading coefficient* of unshaded glass is 1.0 (making this the base condition for comparison).

Shading coefficients shown in Table 2–3.5 are typical for sun-exposed orientations. When the surface being evaluated is in the shade, the coefficient would tend to be about 10–20% higher (meaning the assembly tends to be somewhat *less* efficient in shading diffuse radiation).

Table 2-3.5 Solar Shading

Representative	Typical Shading Coefficient
Interior Blinds and Shades:	
Dark-colored roller shade, half drawn	0.90
Dark-colored roller shade, fully drawn	0.80
Light-colored roller shade, half drawn	0.70
Light-colored roller shade, fully drawn	0.40
Dark gray curtain (6–8 oz/sq yd) fully drawn	0.60
Light gray curtain (6–8 oz/sq yd) fully drawn	0.50
Off-white curtain (6–8 oz/sq yd) fully drawn	0.40
Dark-colored blinds, adjusted to provide full shading	0.75
Light-colored blinds, adjusted to provide full shading	0.55
Metallic aluminum blinds, adjusted to provide full shading	0.45
Absorbing and Reflecting Glass:	
¼-in heat-absorbing glass	0.65
¼-in gray plate glass	0.65
Double sheet: ¼-in heat-absorbing plus ¼-in clear plate	0.65
Double sheet: ¼-in gray plate plus ¼-in clear plate	0.60
¼-in medium-dark tinted glass	0.50
Light gray metallized reflective coating on glass	0.35–0.60
Dark gray metallized reflective coating on glass	0.20–0.35
Adjacent Structures and Landscaping:	
Typical tree which produces light shading on the surface under consideration	0.50–0.60
Typical building or tree which provides dense shading on the surface under consideration	0.20–0.25
External Architectural Elements:	
1. Principally for use when south orientation dominates:	
Continuous overhang, full shade at opening	0.25
Canvas awning, dark or medium color, full shade at opening	0.25
Movable horizontal louvers, adjusted to provide full shading	0.10–0.15
Outside venetian blinds, off-white color, adjusted to provide full shading	0.15
2. Principally for use when east or west orientation dominates:	
Fixed vertical fins or louver grids, full shade at opening	0.30
Movable vertical louvers, adjusted to provide full shading	0.10–0.15

be effective, provided there is some assurance that these elements will be preserved in a useful form.

External Architectural Treatment. Since various wall orientations have differing shading needs, external shielding devices can be architecturally expressive of the regional nature of the building.

FIGURE 2–3.4

EXTERIOR SOLAR SCREENS

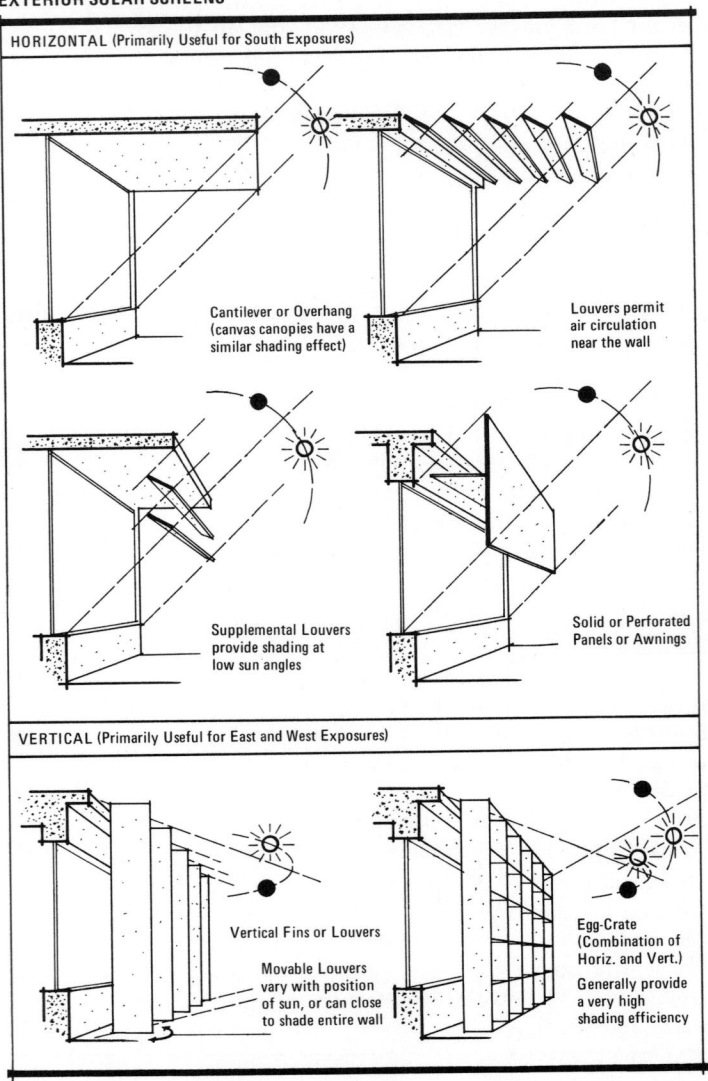

External devices are generally more effective than similar internal devices because energy interception occurs outside of the occupied space, where it can be dissipated with minimum effect on the interior environment.

Architectural shading can take several forms. Shading can be provided by functional elements such as balconies or plan setbacks; it can be provided by structural extensions such as overhanging cantilevers; or it can be provided by secondary structures such as awnings, screens, or fixed louvers.

When winter heat gain is beneficial, these devices should ideally achieve summer shading while permitting winter heat penetration.

Opaque Wall Assemblies. The clear glass opening can also be replaced with opaque wall materials. While this is most effective in the sense of this discussion, it may be objectionable in cases where interior-exterior visual contact is required or desired. However, if the preceding techniques are ineffective or undesirable for other reasons, the opaque assembly may be the most meaningful solution.

Penetration of solar radiation through opaque assemblies

When solar radiation is incident on an opaque enclosure or shell, part of the incident energy is reflected (see Table 2–3.8), a relatively small part is transmitted, and the remainder is absorbed. Since this absorbed energy remains within the assembly until it is re-emitted from either the

Table 2-3.6 Heat Storage Characteristics

Representative Materials		Approximate Time Lag in Heat Flow through Material
Wood	½-in thickness	0 hours 10 minutes
	1-in thickness	0 25
Concrete	2-in thickness	1 hour 5 minutes
	6-in thickness	3 50
	8-in thickness	5 5
	12-in thickness	7 50
Brick	4-in thickness	2 hours 20 minutes
	8-in thickness	5 20
	12-in thickness	8 20
Insulation	2-in thickness	0 hours 40 minutes

NOTE: For composite assemblies, the time lag is approximately additive.
Add an additional ½-hour allowance for a composite assembly that involves two-layer construction.
Add an additional 1-hour allowance for a composite assembly that involves three-layer construction.

exterior or the interior surface, the heat storage properties of the enclosure become a potentially significant consideration in transmission analysis.

This heat storage capacity will tend to produce a time lag—i.e., the interior thermal environment will tend to lag behind the external weather conditions. The magnitude of this time lag will depend on the storage capacity of the enclosure; with the lag generally lengthening as material density or mass increases (see Table 2–3.6).

If the enclosure storage capacity is sufficient, much of the solar heat of the day may be retained and re-emitted to the sky at night. In this situation, the incident solar heat may not impinge significantly on the interior space. If there are negligible internal loads, then, the interior of a massive structure may be relatively cool on a hot day, because the interior surface radiation (MRT) will tend to relate to the cooler temperatures of the previous night.

As a general rule, therefore, massive structures will tend to produce somewhat stable interior conditions; while lightweight structures are, by contrast, much more responsive to short-term solar variations.

Table 2-3.7 Time Lag Requirements to Minimize Internal Thermal Influences During Peak Periods

	Optimum Time Lag Requirements	*Comments*
Roof	10–12 hours	
East wall	12–17 hours	Requirement is not generally practical (too long), design to provide for little or no lag, and accept a near instantaneous load to coincide with cooler morning air temperatures
South wall	6–10 hours	
West wall	5–10 hours	Most significant wall; along with roof, this is the most useful area for massive materials
North wall	5–10 hours	Load is relatively insignificant in the temperate zone

Exterior surface finish Emissivity (relating to absorption or reflection) is a related surface influence. Finishes that *reflect and emit* rather than *absorb and retain* radiation will tend to diminish the flow of energy through the shell.

Part of the incident radiation is visible (light); and in this sense, white finishes are more reflective than black finishes. But part of the radiation involves invisible infrared and ultraviolet, and the reflectivity of this energy will depend more on molecular composition and surface density than on color.

In general, polished metal surfaces will tend to reflect significant portions of the total incident radiation. Painted surfaces, on the other hand, may tend to reflect significant quantities of light energy, but they absorb much of the radiation in the infrared range. However, when exposed to the sky, the diffuse painted surface will tend to lose significant quantities of the absorbed energy; while the metallic surfaces will retain this heat. This accounts for the prominent use of white (rather than metallic) exterior finishes in warm climates.

When exposed to a dark cavity, on the other hand, the white finish will no longer intercept significant heat by reflection and re-radiation, and the metallic surface will now provide a more effective barrier to infrared heat flow. This accounts for the use of reflective metallic sheeting within insulating air spaces. (See Table 2–3.8.)

Table 2-3.8 Reflective Coatings

| | REFLECTANCE | |
Representative Materials and Finishes	Solar Radiation	Cavity Radiation (Thermal Only)
Effective for dissipation of incident radiation to the sky in warm climates:		
White paint	0.71	0.11
White marble	0.54	0.05
Effective for retention of incident radiation in cool climates:		
Limestone	0.43	0.05
Red paint	0.25	0.06
Gray paint	0.25	0.05
Black paint	0.03	0.05
Effective for reflection of invisible thermal radiation within a cavity space:		
Polished silver	0.93	0.98
Polished aluminum	0.85	0.92
Polished copper	0.75	0.85
Aluminum paint	0.45	0.45

NOTE: Unless light-colored materials are self-cleaning or are easily and reliably cleaned at frequent intervals, much lower solar reflectances must be assumed. Generally, reflectances of 0.50 for light-colored walls and 0.10 for dark-colored walls are assumed.

Conductive interaction with the external air mass

The previous discussion of heat storage and interception relates primarily to solar radiation that impinges on opaque assemblies. Gains and losses also occur due to conductive transfers (through the assembly) between interior and exterior air masses.

The intensity of this conductive flow will depend on the temperature difference between the two air masses ($H \sim T_0 - T_1$); and this will, of course, fluctuate with time of day and season of the year. On a warm day, heat gain will take place because conducted energy flows from exterior to interior. At night or on a cold overcast day, heat exchange is negative, and heat loss takes place because conductive energy flow is essentially interior to exterior.

In essence, the transmission process is analogous with the absorption and transmission of moisture by a porous material. Heat is transferred by conduction through successive *layers* of the enclosure until the effect is felt on the opposite surface. Some materials and layers will offer little resistance, absorbing and conducting heat rapidly. Others will exhibit a capability to reflect, store, or otherwise resist the flow.

Surface configuration   Normally, exterior surface temperatures will tend to be higher than the temperature of the surrounding air. As a result, air movement over these exposed surfaces will tend to increase the dissipation of heat from the enclosure shell ($H \sim F_0$).

Table 2-3.9 Surface Conductance (Winter Conditions)

Representative Materials	Wind Velocity			
	10 mph	15 mph	20 mph	25 mph
	($btu/sq\,ft/hr/°F$)			
Stucco	5.5	9.0	11.2	13.6
Brick	4.6	7.1	9.0	10.8
Concrete	4.4	6.8	8.4	10.1
Wood (natural)	3.6	5.8	7.0	8.1
Smooth plaster	3.6	5.5	6.8	8.0
Glass	3.3	5.1	6.2	7.2
Wood (painted)	3.3	5.1	6.2	7.2

NOTE: For interior surfaces that are exposed to reduced air convection, surface conductances can be assumed to be:
1.46 for a wall
1.63 for a ceiling
1.08 for a floor
NOTE: Thermal resistance (R) is the reciprocal of the above conductance rates.

Since this is a negative influence in cold weather, exposed surface area should be minimized in areas where cold temperatures predominate.

When surface cooling is desired, on the other hand, heat dissipation can be increased by utilizing techniques and forms that increase the exposed surface area. These are: (1) the use of curved surfaces such as vaults or domes; (2) the use of uneven surfaces such as corrugated materials or uneven brick coursing; and (3) the use of roughened surface finishes such as stucco or concrete. (See Table 2–3.9.)

Surface conductance will also increase with wind velocity; and although wind direction has a somewhat marginal effect, wind parallel to the surface will produce the highest heat transfer for a given material.

Material resistance   As a general rule, materials that are good conductors of electricity are also good conductors of heat. The converse is also true.

For this discussion, the insulation value of a material can be defined as the resistance (R) of that material to the flow of heat (see Table 2–3.10). Similarly, the resistance value of a structural assembly is the additive sum

Table 2-3.10 Thermal Resistance Rates

Representative *Materials and Assemblies*		*Resistance (R)*	
		btu/hr/sq ft/°F *for 1-in thickness*	
Aluminum, steel, copper		*negligible*	
Wood: Softwood		0.91	
Hardwood		1.25	
Concrete: Composed of sand, gravel, stone		0.08	
Plaster		0.20	
Brick: Common		0.20	
Face brick		0.11	
Glass		0.20	
Insulation: Cellular glass		2.50	
Wood fiber		4.00	
Air space, w/reflective coat	wall	2.64*	
	roof	1.84*	
	floor	5.00**	
Air space, w/o reflective coat	wall	0.92*	
	roof	0.80*	
	floor	1.10**	

*¾–4-inch width.
**1-inch depth.

(R_t) of the individual material resistances (see Figure 2–3.5). The higher the R value, the greater the resistance to conductive heat flow.

As heat flows through a homogeneous material, there is a continual reduction in temperature. Increasing thickness will therefore increase the resistance of the enclosure.

When such increases in thickness are no longer practical or suitable, the resistance of a thicker material can be simulated by utilizing materials of high insulation value (see Figure 2–3.6).

Air spaces Still air is an excellent insulator (see Table 2–3.10). When lightweight structures are desired or required, therefore, the materials in the enclosing shell should be assembled in a manner that will enclose or contain thin layers of air.

The principal insulation value of a vertical air space (in a wall) is achieved at about ¾ in. width. Beyond this width, the improvement curve is essentially flat.

Although an air space is assumed to have zero wind velocity, heat is transmitted across this space by natural convection, by radiation, and by conduction through structural members. In this regard, very significant improvement in insulation value is achieved by adding a metallic reflective coating on the cold side of the cavity space. This coating acts as a reflector of the radiant energy that is emitted across the air space from the warm side.

Estimating conductive heat loss and heat gain Calculations of conductive heat loss and heat gain depend: (1) on the surface area involved; (2) on the transmission coefficient of the assembly (U); and (3) on the temperature differences involved.

For winter heat loss calculations, the difference in temperature between the interior and exterior air masses is used. For estimating summer heat gains, however, *equivalent temperature differences* (Table 2–3.11) are substituted for simple indoor-outdoor temperature differences. (See sample heat loss and heat gain calculations, Figure 2–3.5.) ETD figures represent a means through which temperature differences in the calculations are adjusted to compensate for both air mass differences and solar radiation effects.

Infiltration

Infiltration is an accidental air flow through the enclosing shell; while ventilation is a controlled flow that is intended to maintain the quality of the interior air. In both cases, heat and moisture transfers occur.

FIGURE 2–3.5

BASIC BRICK WALL

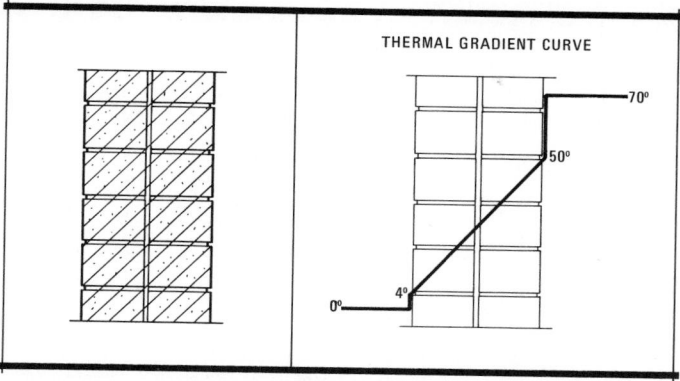

THERMAL GRADIENT CURVE

Thermal Analysis

Thermal Resistance (hr/btu/sq ft/°F)
Assembly: 8-in common brick wall (see Tables 2-3.9, 2-3.10 for data)

Outside surface conductance		
(15 mph wind condition)	1/7.1	= 0.14
8-in common brick	8 × 0.2	= 1.60
Inside surface conductance	1/1.46	= 0.68
	R_{TOTAL}	= 2.42

Coefficient of Heat Transmission (btu/hr/sq ft/°F)
$U = 1/R_T = 1/2.42$
$= 0.414$ but/hr/sq ft/°F

Heat Loss Analysis (Conduction)
Interior temperature: 70°F
Exterior design temperature: 0°F
Wall area being considered: 100 sq ft
$H_C = (Area)(U)(T_I - T_O) = (100)(0.414)(70 - 0)$
$= 2900$ btu/hr loss at 0°

Heat Gain Analysis (Conduction)
Equivalent temp. difference: 30 (from Table 2-3.11)
Wall area being considered: 100 sq ft
$H_C = (Area)(U)(ETD) = (100)(0.414)(30)$
$= 1242$ btu/hr gain

Analysis of Winter Gradients Through Assembly

$\dfrac{0.14}{2.42} \times (70°-0°) = 4°$ drop at outside surface

$\dfrac{1.60}{2.42} \times (70°-0°) = 46°$ drop through brick

$\dfrac{0.68}{2.42} \times (70°-0°) = 20°$ drop at inside surface

Inside surface temperature: 50° (to be used for MRT calculations)

FIGURE 2–3.6

UTILIZATION OF INSULATION

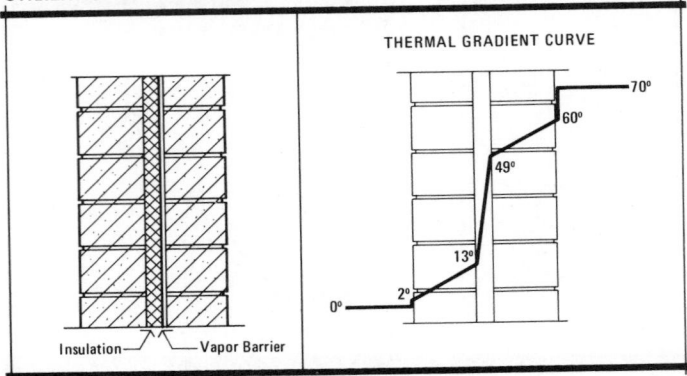

Insulation —/ \— Vapor Barrier

THERMAL GRADIENT CURVE

Thermal Analysis

Thermal Resistance (hr/btu/sq ft/°F)
Assembly: 8-in common brick:
        1-in cellular glass insulation
        (see Table 2-3.9; 2-3.10)

Outside surface conductance (15 mph wind condition)	$1/7.1 = 0.14$
4-in common brick	$4 \times 0.2 = 0.80$
1-in cellular glass insul.	$1 \times 2.50 = 2.50$
4-in common brick	$4 \times 0.2 = 0.80$
Inside surface conductance	$1/1.46 = \underline{0.68}$
	$R_{TOTAL} = 4.92$

Note: 1-in thickness of cellular glass insulation (R = 2.50) is equal in insulation value to an additional 12½″ common brick (R = 0.20)

Coefficient of Heat Transmission (but/hr/sq ft/°F)
$U = 1/R_T = 1/4.92$
   $= 0.204$ btu/hr/sq ft/°F

Analysis of Winter Gradients Through Assembly

$\dfrac{0.14}{4.92} \times (70° - 0°) = 2°$ drop at outside surface

$\dfrac{0.80}{4.92} \times (70° - 0°) = 11°$ drop through outer brick

$\dfrac{2.50}{4.92} \times (70° - 0°) = 36°$ drop through insulation

$\dfrac{0.80}{4.92} \times (70° - 0°) = 11°$ drop through inner brick

$\dfrac{0.68}{4.92} \times (70° - 0°) = 10°$ drop at inside surface

Inside surface temperature: 60° (to be used for MRT calculations)

Dew Point Analysis
  Interior condition: 70°F, 30% R.H.
  Dew point = 37° (from Table 2-3.13)
        Therefore, condensation will occur within the insulation, unless a vapor barrier is placed on the warm side of the dew point.

FIGURE 2–3.7

UTILIZATION OF AIR SPACE

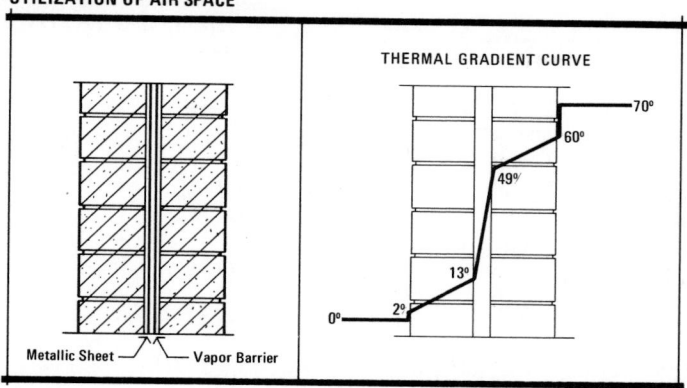

Thermal Analysis

Thermal Resistance (hr/btu/sq ft/°F)
Assembly: 8-in common brick
              1-in air space, w/metallic reflective coating on outer space
              (see Table 2-3.9; 2-3.10)

Outside surface conductance (15 mph wind condition)	$1/7.1 = 0.14$
4-in common brick	$4 \times 0.2 = 0.80$
1-in air space, w/reflective coating	$1 \times 2.64 = 2.64$
4-in common brick	$4 \times 0.2 = 0.80$
Inside surface conductance	$1/1.46 = 0.68$
	$R_{TOTAL} = \overline{5.06}$

Coefficient of Heat Transmission (btu/hr/sq ft/°F)
$$U = 1/_T = 1/5.06$$
$$= 0.198 \text{ btu/hr/sq ft/°F}$$

Analysis of Thermal Gradients Through Assembly

$\dfrac{0.14}{5.06} \times (70°-0°) = 2°$ drop at outside surface

$\dfrac{0.80}{5.06} \times (70°-0°) = 11°$ drop through outer brick

$\dfrac{2.64}{5.06} \times (70°-0°) = 36°$ drop through air space

$\dfrac{0.80}{5.06} \times (70°-0°) = 11°$ drop through inner brick

$\dfrac{0.68}{5.06} \times (70°-0°) = 10°$ drop atinside surface

Inside surface temperature: 60° (to be used for MRT calculations)

Dew Point Analysis
Interior condition: 70°F, 30% RH
Dew point = 37° (from Table 2-3.13)
        Therefore, condensation will occur within the air space, unless
a vapor barrier is placed on the warm side of the dew point.

Table 2-3.11   Equivalent Temperature Differences

	Equivalent Temperature Differences (ETD)	
	surface in sun[1,2]	*surface in shade*[2]
Lightweight Construction:		
Light-colored wall	35	15
Dark-colored wall	50	15
Roof	60	15
Mediumweight Construction:		
Light-colored wall	20	10
Dark-colored wall	30	10
Roof	50	12
Heavyweight Construction:		
Light-colored wall	10	5
Dark-colored wall	15	5
Roof	45	10

[1]Wall figures shown for *sun* conditions apply for the west wall; figures for both east and south walls are approximately 75–85% of the figures shown.
[2]North walls utilize the *shade* figures, as do all orientations at appropriate times of the day.

With ventilation, these transfers are generally localized, and this air may be subject to preconditioning before it enters the occupied space. Infiltration, on the other hand, is a more general and uncontrolled condition that involves random air exchanges through leaks, cracks, and other openings. Control of infiltration, then, is primarily a problem of joint design and construction around window and door openings.

However, the porosity of materials may also be a factor in infiltration. With brick or concrete block, for example, air leakage tends to be relatively high. Application of nonporous films, such as plaster coatings or building paper, will significantly reduce (but not eliminate) this leakage.

In addition to material and construction variables, the actual infiltration air volume (and therefore the heat transfer) will depend on the wind velocity.

Wind effects   Wind will tend to increase static pressure on the outside face on one side of an enclosure shell, while the static pressure on the opposite side of the shell is reduced. As a result, air will tend to flow from exterior to interior through the windward wall (or walls) and from interior to exterior through the opposite wall.

As noted in the previous discussion of surface conductance, heat flow through the wall will also vary somewhat due to wind direction. Maximum surface conductance will occur on walls that are parallel to the direction of wind flow (see Figure 2–3.8).

FIGURE 2–3.8

WIND EFFECTS

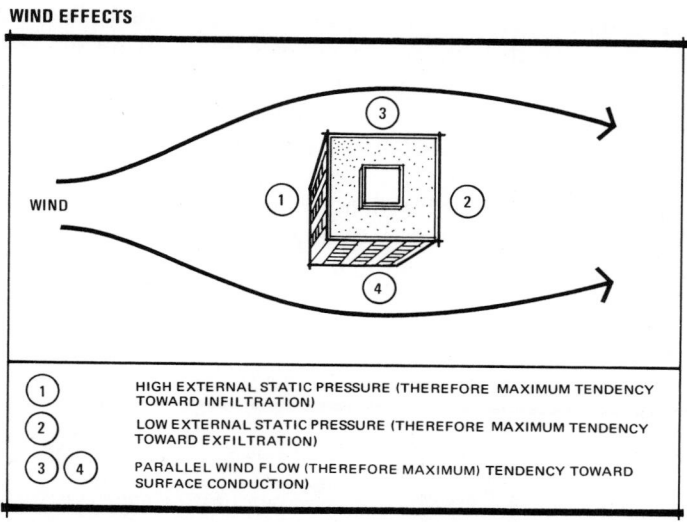

① HIGH EXTERNAL STATIC PRESSURE (THEREFORE MAXIMUM TENDENCY TOWARD INFILTRATION)

② LOW EXTERNAL STATIC PRESSURE (THEREFORE MAXIMUM TENDENCY TOWARD EXFILTRATION)

③④ PARALLEL WIND FLOW (THEREFORE MAXIMUM) TENDENCY TOWARD SURFACE CONDUCTION)

These wind effects tend to become greater as the height above ground level increases. For example, wind velocities at the upper floors of a 40-story building are approximately 1½ times the velocity at ground level. Since static pressure increases approximately as the square of the wind velocity (see Table 2–3.12), such an increase will have a significant effect on the infiltration rate and volume.

Table 2-3.12 Representative Wind Pressures

Wind Velocity		Static Pressure
miles per hour	feet per minute	inches of water
5	440	0.015
10	880	0.05
15	1320	0.10
20	1760	0.18
25	2200	0.29
30	2640	0.42

Stack effects   Because of the natural tendency for warm air to rise, significant vertical variations in interior static pressure are induced, and these tend to produce a pronounced *chimney effect* in tall buildings. This will induce infiltration on all walls at the lower floor levels, while inducing exfiltration (outflow) at the upper floors. Near the middle floors, there is a *neutral zone* at which there is little pressure gradient for air flow either way (except for the previously discussed wind effects).

This *stack effect* is minimized by sealing all openings between floors (such as stairways, elevator shafts, and mechanical shafts).

Pressurization   Since infiltration of outdoor air will carry dust and contaminants as well as inducing locally variable heat gains and losses, conscious effort is often made to minimize this effect by *pressurizing* the

FIGURE 2–3.9

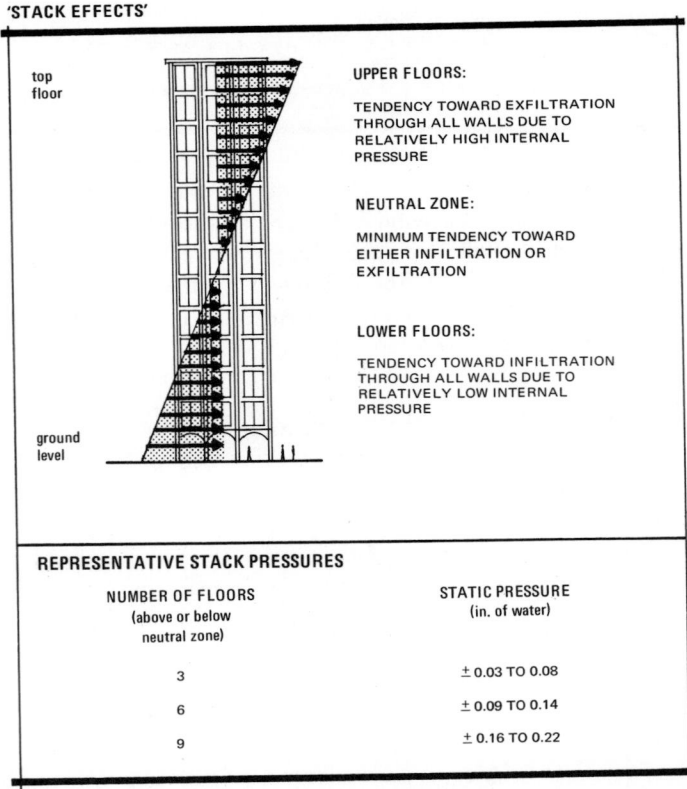

'STACK EFFECTS'

NUMBER OF FLOORS (above or below neutral zone)	STATIC PRESSURE (in. of water)
3	± 0.03 TO 0.08
6	± 0.09 TO 0.14
9	± 0.16 TO 0.22

REPRESENTATIVE STACK PRESSURES

building interior. This is accomplished by reducing the ventilation exhaust capacity relative to the ventilation intake or supply capacity, thereby causing a rise in interior static pressure.

This applied force will tend to induce exfiltration through all cracks, openings, and porous materials. For this reason, it must be of sufficient pressure to balance or counteract the previously discussed infiltration that results from wind or stack effects.

Vapor barriers Condensation of water vapor will occur within wall, roof, or floor assemblies at the temperature plane where seeping air becomes saturated. This temperature plane is defined by the dew point temperature of the air mass on the warmer side of the enclosure shell (see Figure 2–3.6 and related *dew point analysis*).

Since moisture can cause physical damage and deterioration within the assembly, it is necessary to minimize this effect. One method involves the reduction of humidity on the warm side of the barrier. But relative humidity levels must remain moderately high to meet human comfort standards. To prevent vapor transmission and subsequent condensation within the assembly, therefore, an impervious vapor barrier should be placed between the theoretical dew point plane (developed for the maximum design temperature gradient) and the moisture-laden warmer air mass.

The nearer the vapor barrier is to the moisture source, the greater will be the moisture limit that can be tolerated in the warm air mass. Ideally, then, the vapor barrier should be placed near the inner surface in cold climates and near the outer surface in hot, humid regions. Since the greatest temperature and absolute humidity differences between air masses occurs during the winter months in north temperate regions, the vapor barrier is generally placed near the inner surface in these climate zones.

Table 2-3.13 Dew Point Temperatures

Interior Environmental Temperature	Relative Humidity					
	30%	40%	50%	60%	70%	80%
60°F	29°	35°	41°	46°	50°	54°
65°F	33°	40°	46°	51°	55°	58°
70°F	37°	45°	51°	56°	60°	64°
75°F	41°	49°	55°	60°	65°	69°
80°F	46°	53°	60°	65°	69°	73°

NOTE: For identification of dew point locations, see "analysis of winter gradients through assemblies," (Figures 2-3.6, 2-3.7).

Internal Heat Gains

Not too many years ago, builders of human enclosures were dealing with situations where the heat generated by the occupants themselves was practically the only normal source of internal heat gain (exclusive of heating devices). The major problem in thermal design was to control heat excesses or deficiencies related to weather or climatic conditions.

But this situation is changing; for today, it is found that internally-generated heat may actually be the more prevalent thermal influence during many of the normal use-hours of some building types.

It is increasingly relevant, therefore, that internal heat gains be evaluated for the possibility that they will produce conditions that require modification of conventional ideas of orientation and utilization of perimeter building materials. Furthermore, methods of internal heat manipulation and control become increasingly significant.

Heat gains associated with occupancy and ventilation

One category of significant internal heat gain is associated with the human occupancy requirements.

Part of this heat is associated with the dissipation of occupant heat itself—heat that is generated by the internal metabolic processes of the body (see Table 2–3.14). For individuals, this heat tends to be somewhat minor when compared with other conduction and radiation transfers. But occupancy can become a significant gain factor when the density of occupancy increases. This category, then, is particularly meaningful for enclosures that are intended for congregation and assembly activities.

Table 2-3.14   Heat Gains Associated With Occupancy

Conditions for Average Individuals	Sensible Heat	Latent Heat	Total Heat
Seated at rest	180	150	330
Seated, office work	200	250	450
Standing, light work; or walking slowly	200	300	500
Light work	220	530	750
Moderately heavy work; or walking at a brisk rate	300	700	1000
Heavy work	465	985	1450

NOTE: Above figures are based on an average group of men, women, and children.
Female: Approx. 85% of the heat gain for an adult male.
Children: Approx. 75% of the heat gain for an adult male.

(Adapted: Ashrae, 1961)

A second category of occupancy heat gain is associated with the ventilation air that is introduced for hygienic purposes (see Table 1–3.6). Since exterior air replaces interior air, this process may introduce adverse temperature and humidity conditions into the interior space.

Heat gains associated with equipment

Another category of internal heat gains is associated with equipment that is installed for specific activity needs. For example, the increasingly prevalent use of electric typewriters, calculators and other clerical machines, manufacturing devices, and heat-producing educational and communication aids all add heat to the interior space.

In estimating this load, allowances of 1 watt/sq ft or more should be included as a general distributed load; with special concentrated loads as necessary to accommodate special needs for computer operation, cooking processes, etc.

In analysing the interior thermal balance, it should be noted that some of this internally-generated heat is in a form and in a location that permits partial isolation and control. For example, many large machine and computer elements are in somewhat remote or isolated areas. A portion of this heat can therefore be harnessed through manipulation of air circulation or fluid flow; techniques which conduct considerable quantities of heat away before it can penetrate sufficiently to become a direct influence on occupant comfort.

Heat gains associated with electric lighting

The use of electric lighting produces another important internal heat gain. This applies both in situations where the lighting is intended to implement visual performance in work spaces, and where spatial systems provide environmental lighting for orientation and circulation, assembly, or relaxation (see Chapter 1–1). In both cases, the heat generated by these systems may become a significant influence in the thermal environment.

The increasing significance of this heat in recent years is suggested in a comparative thermal analysis of a representative 5-story building (100 ft × 100 ft per floor, with 33% of each wall area composed of glass).

It was found that an internal lighting load of 3 watts/sq ft accounts for approximately 22% of the total refrigeration load; and the internal heat gain (including a distributed equipment and occupancy allowance) balances the building losses when the outdoor temperature is 48°F or above. As this single internal load (lighting) is doubled to 6 watts/sq ft, and all other variables remain unchanged, the lighting load now accounts for approximately 36% of the total refrigeration load; and the internal heat gain balances the building losses when the outdoor temperature is 25°F or above. Again doubling the lighting load to 12 watts/sq ft, while other thermal loads remain unchanged, the lighting now accounts for 54% of the total refrigeration load; and the internal heat gain balances the building losses when the outdoor temperature is −20°F (below zero) or above.

Control of electric lighting heat The availability of this internal energy may obviously be a decisive consideration in thermal system design. If this energy remains randomly distributed and uncontrolled, it can become a somewhat chaotic thermal influence. But if this energy can be effectively harnessed, it can be utilized or exhausted as necessary to respond to demands imposed by the outdoor environment. Of particular interest, then, is the somewhat isolated location of many lighting devices and their character as heat traps.

In initially evaluating the thermal influence of the electric lighting system, the designer must estimate the action of the electric lamp as a source of heat, as well as a source of light. Heat is emitted at the rate of 3.415 btu/watt-hour of energy consumed. From this is derived the conventional rule of thumb that one ton of cooling capacity is required to remove the heat from approximately 3500 watts of electrical load.

But the lamp is generally placed in an enclosure of some type (a fixture or a ceiling cavity), and this begins to alter the distribution of heat. When the lamp is first turned on, the ballast heat (if there is a ballast), the heat of the bulb itself, and the heat conducted to or absorbed by the air immediately around the bulb is all trapped within the fixture or cavity. This heat may not remain in this location for long; but for a short period of time, it is confined in an enclosure that is somewhat remote from the immediate location of the occupant.

Furthermore, it should be recognized that although part of the light is emitted from the luminaire, part is trapped and absorbed *as heat* by louvers, reflector surfaces, and diffusing panels. The invisible infrared and ultraviolet energies follow a similar path. The precise distribution of these wavelengths depends on the fixture materials (see Table 2–3.15). But again, part of the energy is emitted immediately into the room, while the remainder is absorbed *as heat* in the fixture housing or cavity.

Isolation of heat within the luminaire. By utilizing the luminaire or cavity as a *heat trap,* a substantial quantity of lamp and fixture heat can be brought under control by moving water or return air through the unit. Figure 2–3.10 shows that up to 80% of the total lighting heat is controlled in a prototype system that cycled return air from the room past

FIGURE 2–3.10

ISOLATION OF LIGHTING HEAT

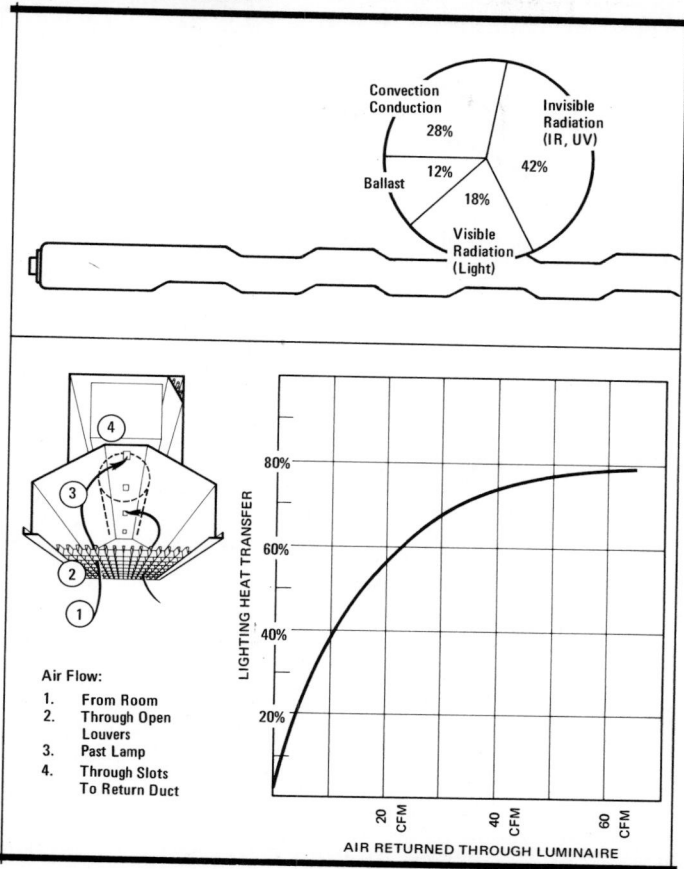

a 1500 ma fluorescent lamp; then through slots in the fixture into a collection duct.

Since most of the heat is easily controlled in the duct system, this technique produces a very significant reduction in the influence of lighting heat in the occupied space itself (see Figure 2–3.11).

Luminaire air and surface temperatures. In normal operation, the heat that is retained within an unventilated luminaire tends to build up, until fixture surfaces and surrounding ceiling areas become secondary heat sources. This heat then moves into the occupied part of the room by convection and re-radiation.

FIGURE 2–3.11

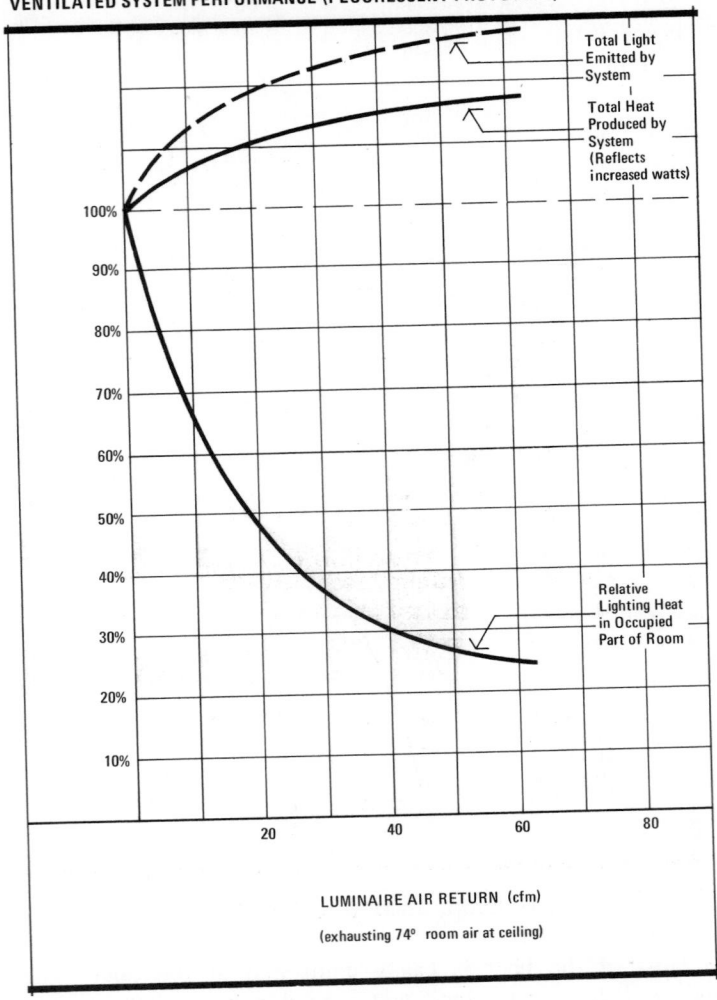

VENTILATED SYSTEM PERFORMANCE (FLUORESCENT PROTOTYPE)

In this regard, surface temperatures in many typical unventilated luminaires often range above 100°F, to as high as 140°F. Where these elements cover a significant portion of the ceiling, therefore, they tend to
perform in a manner similar to a *panel heating* system.

In contrast to this, Figure 2–3.12 indicates the effect of ventilation on
luminaire air temperatures in the representative prototype system. Figure
2–3.13 further indicates the declining luminaire surface temperatures. This
latter reduction is particularly significant in low ceiling rooms.

FIGURE 2–3.12

LAMP AMBIENT CONDITIONS

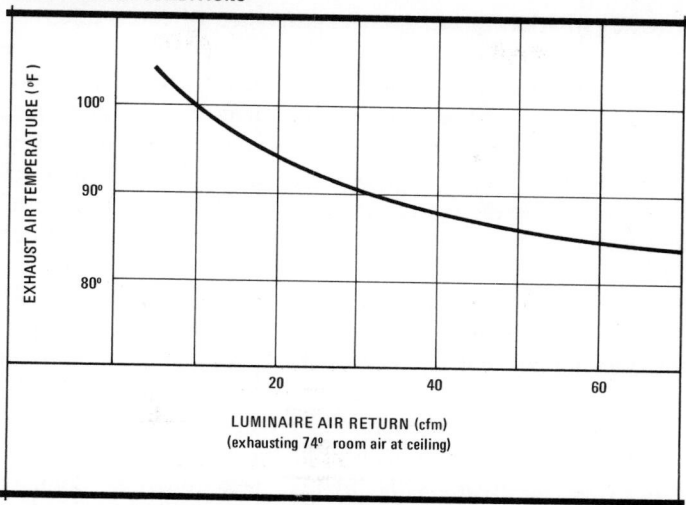

LUMINAIRE AIR RETURN (cfm)
(exhausting 74° room air at ceiling)

FIGURE 2–3.13

LUMINAIRE SURFACE TEMPERATURE

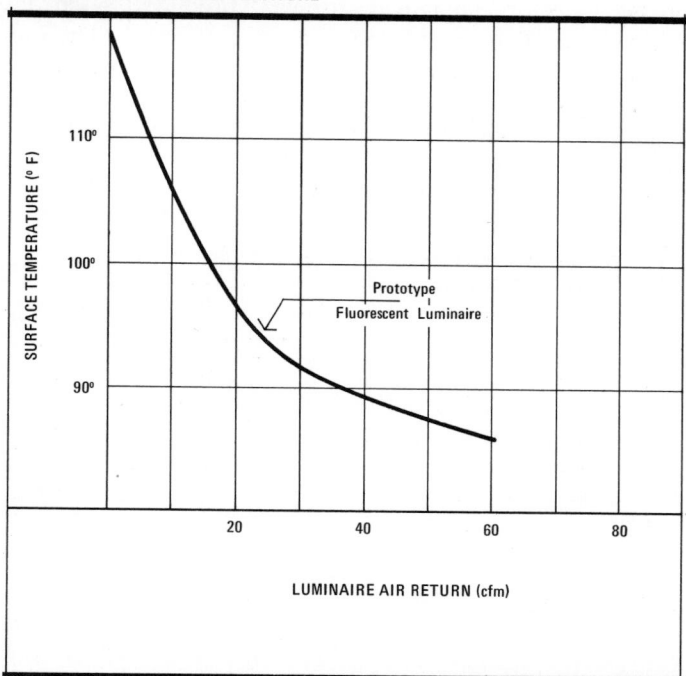

LUMINAIRE AIR RETURN (cfm)

Selection of luminaire materials.   At the normal operating temperatures associated with fluorescent lamps, the glass bulb emits a substantial quantity of infrared radiation in the wavelength range of 5–20 microns.

Related to this is the fact that surface materials have varying reactions to this wavelength band (see Table 2–3.15). So the distribution of this energy will depend on the shielding and reflector materials used in the system. Polished aluminum, and to a lesser extent diffuse anodized aluminum, reflects much of the energy into the space to be lighted (i.e., the infrared tends to follow the light). On the other hand, synthetic enamel on metal will absorb most of the energy (i.e., infrared heat is retained in the luminaire).

A transmitting material such as glass or plastic will also affect the distribution of luminaire energy. These materials permit a high percentage of the visible radiation (light) to enter the space; while most of them are essentially opaque to much of the invisible radiation emitted by the fluorescent lamp. This latter energy is either reflected or absorbed by the panel, and is initially confined within the luminaire or ceiling cavity.

If it is desirable to dissipate the lighting heat into the room, then, the luminaire should utilize polished or diffuse anodized reflector and louver surfaces. On the other hand, if it is desirable to isolate heat within the luminaire, the heat control efficiency is improved by applying white enamel on the reflector and louver surfaces (producing good absorption of infrared, while maintaining efficient reflection of light). The louver grid can also be selected or designed to present a greater absorbing surface by using a smaller cell size and a deeper cell depth. And where suitable, a plastic or glass cover panel will contribute to the heat collection efficiency of the luminaire.

Table 2-3.15 Material and Finish Response to Infrared

	Reflectance at Indicated Wavelength					
	4μ	*7μ*	*10μ*	*12μ*	*15μ*	*20μ*
Polished aluminum	.92	.96	.98	.98	—	—
Diffuse anodized aluminum	.12	.21	.09	.08	.06	.06
Synthetic enamel on steel	.03	.01	.01	—	—	—
Porcelain on steel	.05	.03	.09	.05	.06	.13

Channeling of return air.   Effective removal and control of lighting heat by exhaust air requires a low velocity air path that moves past the lamp.

This circulation *immediately past the light source* is an important factor in the efficient collection of heat.

This point can be demonstrated most dramatically with an incandescent lamp, although the general relationship of air flow and heat collection applies equally when a fluorescent lamp is involved. The maximum bulb temperature of a 100-watt filament lamp operating in 77°F air is 260°F; but the air temperature 1 in. away is less than 15° above the general ambient air. At 3 in., the temperature gradient is insignificant (see Figure 2–3.14).

FIGURE 2–3.14

CHANNELING OF RETURN AIR

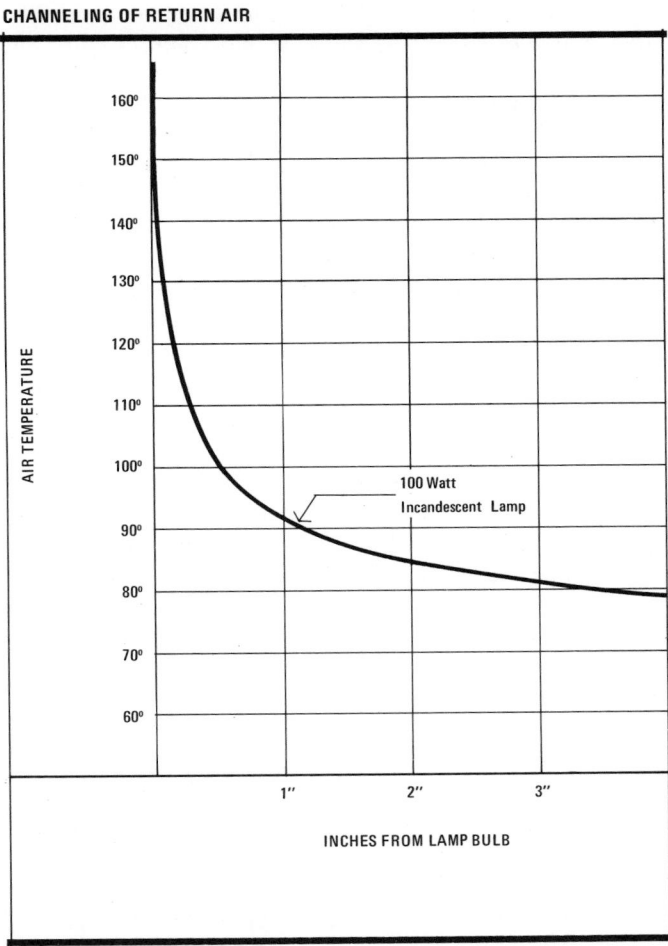

100 Watt
Incandescent Lamp

AIR TEMPERATURE

160°
150°
140°
130°
120°
110°
100°
90°
80°
70°
60°

1"   2"   3"

INCHES FROM LAMP BULB

Because exhaust velocities are low, then, regularly spaced vents or slots should be placed as near to the bulb as possible in order to channel a maximum quantity of this low velocity air through the immediate environment of the lamp itself.

At the same time, it is equally important that this air be channeled to move along the metal fixture surfaces, to pick up radiant energy that has been intercepted and absorbed by these surfaces. The regular spacing of air slots (rather than a single hole) will help this aspect by facilitating more uniform air motion through the system.

Performance of ventilated fluorescent lighting systems.   The prevention of uncontrolled heat build-up within the luminaire, using low velocity return air as previously discussed, produces ambient conditions more favorable for highly efficient operation of fluorescent lamps. Light output and efficiency decline measurably when the ambient air temperature (around the lamp) exceeds 80°F, and ventilation tends to maintain the ambient near this optimum.

However, care is necessary to prevent negative influences. The flow of high velocity *supply air* must be rigidly controlled so as not to become an influence on the fluorescent lamp. Irregular cooling or over-cooling produce noticeable shifts in color and brightness.

Ballast performance is also responsive to temperature, so case temperatures should not exceed 194°F. With excessive temperature, the capacitor will fail, causing the transformer to overheat and causing eventual failure. So ventilated systems are generally beneficial in maintaining safe operating temperatures; and usually, this increases ballast life.

Performance of ventilated incandescent lighting systems.   Significant incandescent lighting heat can also be isolated and controlled. In the prototype systems indicated in Figures 2–3.15 and 2–3.16, three standard reflector-type lamps are evaluated.

In each case, the filament itself operates at a very high temperature (4000–5000°F). So any normal change in the temperature of the air surrounding the bulb is relatively insignificant and will not affect filament temperature. Since filament temperature is neither increased nor decreased, there is no effect on lamp life, light output, or color.

While the efficiency and operation of a filament lamp is relatively unaffected by temperature, however, the high concentrations of heat which are apt to occur with incandescent lamps may adversely affect lamp and fixture materials, causing bulb glass breakage, base seal failures, or socket and wiring deterioration.

In all cases, then (fluorescent and incandescent), an improved tempera-

FIGURE 2–3.15

VENTILATED SYSTEM PERFORMANCE (SINGLE INCANDESCENT)

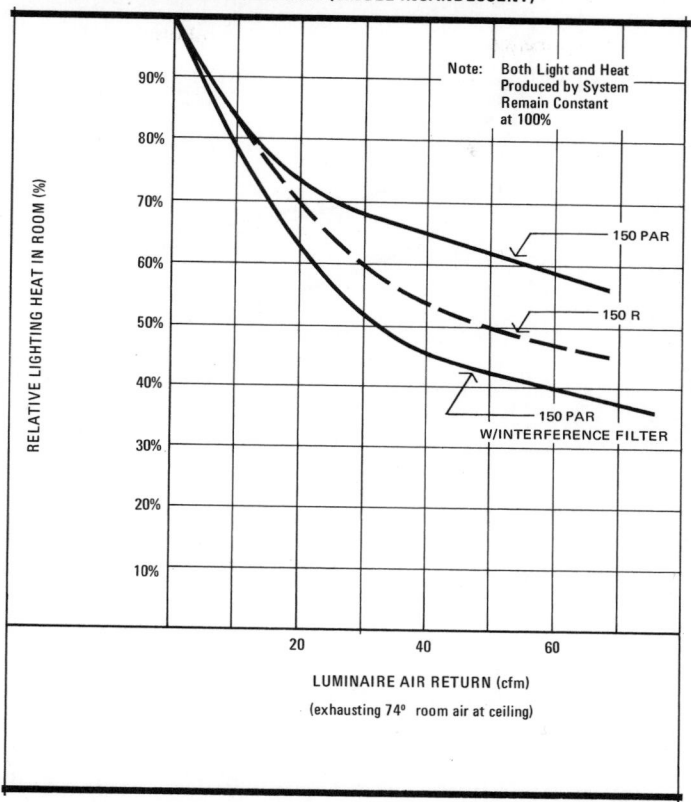

ture condition minimizes the possibility of adverse effects on the materials and components of the electrical system.

Fluid control. The relative heat transferability of a given volume of water is more than 200 times that of the same volume of air. The implications of this dramatic ratio on building space requirements is one reason for the widespread interest in water control systems.

Panel control of luminaire heat is accomplished with tubing attached to the fixture surfaces, or portions of the luminaire can be formed from sheet metal which contains an integral tubing pattern. This technique is particularly effective in controlling energy that is absorbed as heat by metal fixture surfaces. So material and finish selection in the design of these fixtures is based, in part, on the need for surfaces that reflect maxi-

FIGURE 2–3.16

VENTILATED SYSTEM PERFORMANCE (4-LAMP WALL LTG.)

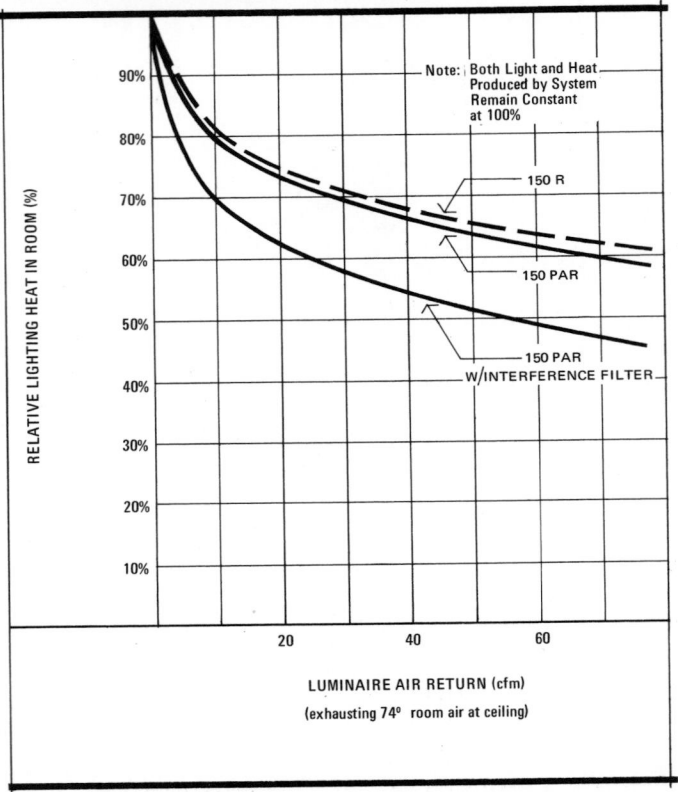

mum quantities of visible energy (light), while absorbing most of the invisible wavelengths (ultraviolet, infrared). This generally leads to the use of white enamel finishes (see Table 2–3.15).

The use of water-cooled panels reduces but does not eliminate the air circulation requirements in the room. Some air is still required for humidity control and for minimum ventilation requirements. Nevertheless, the reductions in air volume will significantly reduce the number and size of air ducts.

Quantitative Estimates: Enclosure Heat Loss and Heat Gain

When estimating the comprehensive response of buildings to internal and external thermal influences, the formats shown in Tables 2–3.16 and 2–3.17 are useful for summary purposes.

Table 2-3.16 Estimation of Enclosure Heat Losses

	Formulas	References
Transmission Losses:		
Opaque and glass walls	$H_c = (U_w)(area_w)(T_1 - T_o)$	see Figure 2-3.5
Roof	$H_c = (U_r)(area_r)(T_1 - T_o)$	
Floor slab	$H_c = (area_r)(2 \text{ btu/sq ft})$	
Slab edge	$H_c = (\text{edge length})(40 \text{ btu/ft})$	see note 1
Basement wall (if any)	$H_c = (area_{bw})(4 \text{ btu/sq ft})$	
Ventilation or Infiltration Losses:		
	$H_v = 0.018(\text{air volume})(T_1 - T_o)$	see note 2
Total		
This figure to be used for estimating the size of both heating elements and air handling elements		

[1] Slab, insulated but unheated ($T_o = 0^\circ$): assume 40 btu/ft/hour edge loss. Slab, insulated and heated ($T_o = 0^\circ$): assume 65 btu/ft/hour edge loss.

[2] For pressurized buildings, the hourly ventilation rate should be used in calculating *air volume* (see Table 1-3.6).
For nonpressurized buildings, use the following for preliminary estimates:

no openings		½ air change/hour
openings:	1 side	1 air change/hour
	2 sides	1½ air changes/hour
	3 or 4 sides	2 air changes/hour

Table 2-3.17 Estimation of Enclosure Heat Gains

	Formulas	References
Transmission Gains:		
Opaque walls	$H_c = (U_w)(area_w)(ETD_w)$	see Figure 2-3.5
Roof	$H_c = (U_r)(area_r)(ETD_r)$	Table 2-3.11
Glass	$H_c = (U_g)(area_g)(T_1 - T_o)$	
Solar Gains:		
Wall openings north	$H_s = (area_g)(solar\ gain)(shading\ coef.)$	see Tables 2-3.1
east		2-3.2
south		2-3.3
west		2-3.5
Roof openings (if any)	Same formula	
Internal gains:		
Occupancy	$H_o = (no./occupants)(sensible\ metabolic\ rate)$	see Table 2-3.14
Lighting	$H_1 = (watts/sq\ ft)(area_r)(3.415\ btu/hr)$	(sensible)
Electrical and mechanical equipment allowance	$H_e = (watts/sq\ ft)(area_r)(3.415\ btu/hr)$	
Subtotal This figure to be used for estimating the size of air handling elements		
Ventilation and humidity allowance:	Utilize psychrometric chart (enthalpy) to determine sensible and latent loads; *or* for preliminary estimates, use an adder of 30% × subtotal	see Tables 1-3.6 2-3.14 (latent)
Total This figure to be used for estimating the size of refrigeration elements		

Balancing Systems

In the temperate zone, buildings tend to be periodically exposed to an extreme range of climatic influences; ranging from extreme cold, to extreme hot-arid, to extreme hot-humid conditions. The optimum enclosure response to some of these thermal conditions is a compact, insulated building that will minimize surface exposure to the hostile external environment. At other times, the optimum response is an open structure that will facilitate the penetration of cooling breezes.

This inconsistency in climatic demands makes it extremely difficult to develop a single optimum building shell that will satisfy all of the weather variations that will occur throughout the year. For this reason, a variety of energy-balancing devices have been developed to compensate for the periodic discrepancies that will inevitably occur when attempting to optimize the response of the building to varying climatic influences.

Heat-generating devices

When there is a need to provide interior heat to compensate for heat losses in cold weather, there are several categories of systems for this purpose. Some utilize combustion processes, while others are noncombustion systems.

Among the *noncombustion types* are two that will be discussed later under the heading of "Cooling Devices." These are combination heating-cooling systems that are capable of providing environmental heat during periods of heat loss. The thermoelectric unit is one such device; one which relies on the ability to reverse a flow of direct current in order to produce either heating or cooling. A second system is the *heat pump,* which relies on the ability to utilize both condenser heating capacity and evaporator cooling in response to varying demands.

Recent attention has also been focused on the significance of normal interior heat gains (such as lighting and equipment loads). The potential for controlling this heat in a predictable manner has been previously discussed in this chapter.

A third category under the general classification of noncombustion systems involves devices that utilize electric resistance heating.

Regarding the *combustion-type* devices, these are generally processes through which *natural fuels* (such as wood or peat) and *fossil fuels* (such as gas, oil, or coal) are used to heat water, to generate steam, or to heat air directly.

Fuels and equipment requirements   Geographic proximity to sources of fuel supply affects transportation costs. For this reason, proximity will often be a decisive factor in the selection of appropriate equipment to balance building heat losses in cold weather.

However, space requirements for generating equipment, for fuel storage, for chimneys, and for distribution devices may become an offsetting consideration. So equipment characteristics and equipment space requirements may also become decisive economic or design factors.

In the comparative chart (Table 2–3.18), it will be noted that fuel costs tend to be low for combustion-type processes, while equipment needs will be relatively extensive. For this reason, initial equipment costs and equipment maintenance costs will tend to be high.

On the other hand, most noncombustion processes present less extensive generating requirements. Chimneys and fuel storage are eliminated; space for equipment is often reduced; and initial equipment costs may be somewhat lower. But since most of these systems rely on some form of electricity, fuel costs tend to be relatively high.

The correct or optimum selection from among these alternatives will depend on the requirements and limitations of the specific project.

The combustion process and distribution of the heating medium.   As fuels are burned in combustion-type systems, oxygen is required to support the combustion. Since air contains only about 20% oxygen, very significant volumes of air must be brought to the immediate proximity of the burner (where the fuel is located).

But in addition to heat, combustion also produces carbon wastes. In this sense, inadequate flow of combustion air produces an uneconomic use of fuel. This is a *sooty* operation, and it produces excessive quantities of carbon monoxide. For these reasons, optimum quantities of air and proper flow through the burner are decisive requirements.

The chimney is a basic element in controlling this flow of combustion air and gases. It actually performs two functions: (1) inducing a flow of air (draft) through the combustion chamber, and (2) carrying off the wastes (smoke and flue gases).

In combustion-type systems, then, the combustion process itself defines one route-of-flow through the heat generation system (see Figure 2–3.17). Heated flue gases tend to rise through the chimney by natural or forced convection. This process not only evacuates undersirable gases, but it also induces a flow of fresh air into the combustion chamber to replace the exhausting gases. This constant flow of fresh air supplies the oxygen required to sustain the combustion process.

Table 2-3.18 Comparison of Heating Systems

	Fuel Cost	Generation Equipment Requirements				Distribution Equipment Requirements		
		Condenser-Evaporator Equipment	Furnace or Boiler	Chimney Action	Fuel Storage	Ducts or Pipes for Supply and Return	Fans or Pumps	Convectors or Panels
Noncombustion Heat Sources:								
1. Utilization of condenser heat (including heat pumps)	high	yes	—	—	—	piping for water or refrigerant	yes	yes
2. Thermoelectric	high	—	—	—	—	electric wiring	—	yes
3. Utilization of interior waste heat	—	—	—	—	—	air or water cycle	yes	yes
4. Electric resistance units	high	—	—	—	—	electric wiring	—	yes
Combustion-Type Heat Sources:								
1. Natural fuels (wood, peat)	low	—	yes	yes	yes	air or water cycle	yes	yes
2. Fossil fuels coal	low	—	yes	yes	yes	air or water cycle	yes	yes
gas	med.							
oil	med.							

FIGURE 2–3.17

HEAT GENERATION (COMBUSTION)

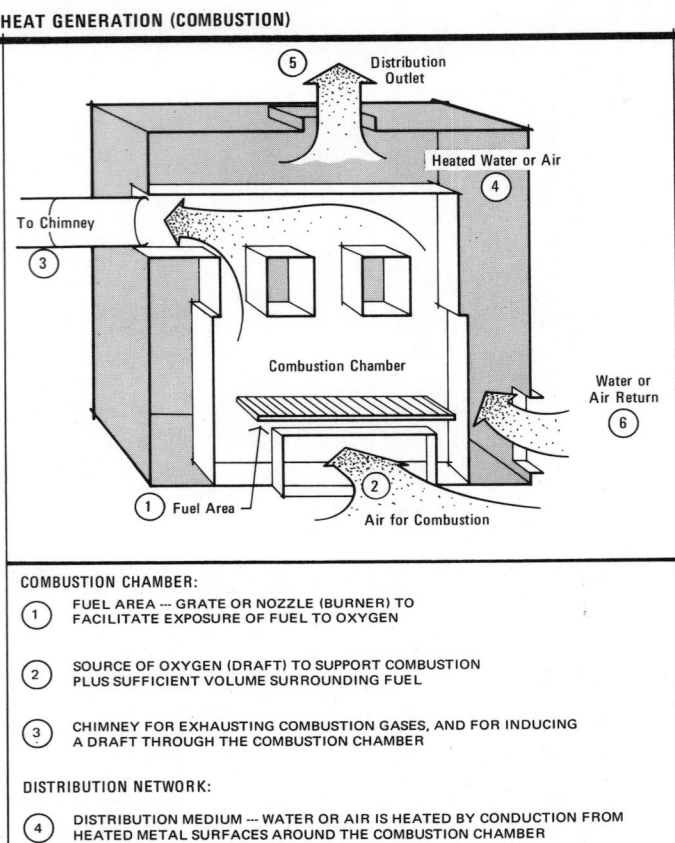

COMBUSTION CHAMBER:

(1) FUEL AREA — GRATE OR NOZZLE (BURNER) TO
FACILITATE EXPOSURE OF FUEL TO OXYGEN

(2) SOURCE OF OXYGEN (DRAFT) TO SUPPORT COMBUSTION
PLUS SUFFICIENT VOLUME SURROUNDING FUEL

(3) CHIMNEY FOR EXHAUSTING COMBUSTION GASES, AND FOR INDUCING
A DRAFT THROUGH THE COMBUSTION CHAMBER

DISTRIBUTION NETWORK:

(4) DISTRIBUTION MEDIUM — WATER OR AIR IS HEATED BY CONDUCTION FROM
HEATED METAL SURFACES AROUND THE COMBUSTION CHAMBER

(5) DISTRIBUTION OUTLET (UTILIZING THE NATURAL TENDENCY
FOR WARMED AIR OR WARMED WATER TO RISE)

(6) DISTRIBUTION RETURN INLET (UTILIZING THE NATURAL TENDENCY
FOR COOLED AIR OR COOLED WATER TO SEEK A LOWER LEVEL)

The second route-of-flow is completely separate and relates to the heat-distribution medium. This medium is warm air (for a furnace) or hot water or steam (for a boiler). It is initially collected around the perimeter of the combustion chamber, where it receives heat by conduction through the metal chamber enclosure. As this medium is heated, then, normal gravity convection will begin to induce a natural circulation, with the heated air or water rising through a duct or pipe system.

After the circulating medium discharges its heat to an occupied space, the cooled medium will tend to seek a lower level. A system of return ducts or pipes is provided for this; and this completes the cycle back to the perimeter of the combustion chamber, where the reheating process begins again.

This distribution process can utilize natural gravity flow, as described; or fans and pumps can be introduced to induce a more variable or precise rate of flow (see Figure 2–3.18).

FIGURE 2–3.18

BASIC HEATING SYSTEMS

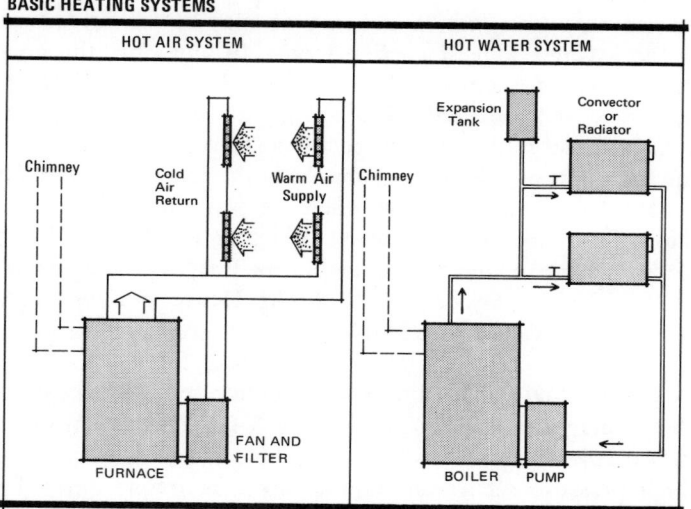

Cooling devices

When excessive heat or humidity accumulate in the interior space, this can be neutralized with refrigeration.

In previous years, cooling action was accomplished by fusion processes. This utilized changes of state such as ice to water (solid to liquid). Heat from the overheated space was absorbed at a rate of 144 btu in the conversion of each pound of ice to water (144 btu/lb is the *heat of fusion* for water).

In recent years, however, mechanical compression and absorption processes have utilized the *heat of fluid* and *heat of evaporation* to accomplish a cooling action. Again, there may be a change of state which causes the refrigerant to absorb heat from the interior air mass. If water is the refrigerant, for example, the purpose of the system may be to create conditions in which the water will evaporate. When this is done, the *heat of evaporation* will absorb approximately 970 btu from the interior air mass in the conversion of each pound of water to vapor.

Since the useful refrigeration effect depends on the quantity of heat absorbed in the change of state, the evaporation process is more efficient than the earlier-used fusion processes. Furthermore, the ability to transport, store, and process liquids is generally more effective than it is for solid refrigerants.

In order to complete the refrigeration cycle, then, the refrigerant must be reconverted to the original state. For evaporative processes, this is generally accomplished by condensing the refrigerant vapor and expelling the sorted heat to a *heat sink*. This heat sink may be the outdoor air mass or it may be a water well or cooling tower. In any case, the greater the temperature difference between the condenser heat and the temperature of the heat sink, the more effective will be the ability of the system to dissipate its heat and return the refrigerant to the original state.

The common unit of refrigerant capacity is the *ton of refrigeration*. This unit is a carry-over from the time when ice was the principal means of refrigeration. Since the heat of fusion of water is 144 btu/lb, the heat absorbed by one ton of ice (2000 lb) in melting over a 24-hour period would be 288,000 btu/day or 12,000 btu/hr. Hence, *one ton* of refrigeration capacity is equivalent to 12,000 btu/hr.

Compression devices for cooling   Cooling by mechanical compression is shown diagrammatically in Figure 2–3.19. This action involves the following processes:

(1) A high pressure refrigerant liquid (such as freon) flows from the compression chamber to a lower pressure chamber called the evaporator.

(2) Cooling action occurs due to evaporation of the refrigerant in the low pressure evaporator chamber; with the heat of evaporation required for this process being drawn from the interior air mass.

(3) The resulting low pressure refrigerant vapor from the evaporator passes to the compression chamber, where mechanical compression of this gas increases the pressure and causes it to reach a saturation condition in the condenser.

(4) With the mechanical compression, the refrigerant vapor returns to liquid form in the condenser; with the stored heat being rejected to outdoor air or to cooling water. (The cooling water then passes into a well, to a cooling tower, or to an air-cooled condenser.)

(5) The resultant high pressure liquid flows to the evaporator, and the cycle repeats.

FIGURE 2–3.19

THE COMPRESSION REFRIGERATION CYCLE

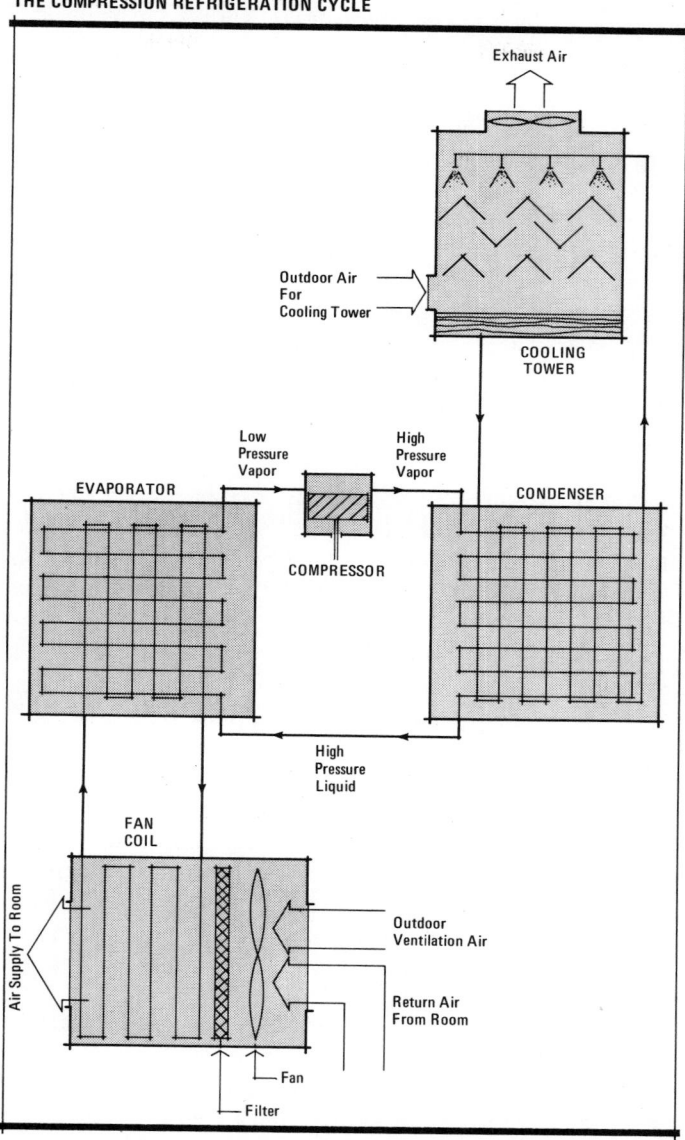

The heat pump cycle. When the compression cycle is made reversible, so that the relative positions of interior and exterior air masses can be interchanged, the system is called a *heat pump*. This principle provides a system that can either provide interior *cooling* (by using the evaporator capability) or interior *heating* (by using the condenser capability).

Absorption devices for cooling   When refrigeration is accomplished by mechanical compression, the input energy is shaft work. This is relatively expensive because, in compression, the vapor undergoes a large change in specific volume. If it were possible to raise the pressure of the refrigerant without appreciably altering its volume, the required work (i.e., input energy) would be reduced.

The absorption refrigeration cycle accomplishes this with absorption of the refrigerant vapor by a liquid (such as water). Although a very significant reduction in shaft work is accomplished by this method (i.e., fewer moving parts), heat input must be increased several times over that required for the mechanical compression cycle. For this reason, an inexpensive source of heat is required in order to make the absorption cycle attractive.

As in the mechanical compression system, the useful refrigeration effect of an absorption system depends on the change of state (or enthalpy change) within the evaporator. However, several additional components are required to perform the function of the compressor-condenser combination (see Figure 2–3.20). The functions of these components are as follows:

(1) An absorber contains a salt solution that absorbs water vapor from the evaporator This action increases the tendency of water to evaporate in the relatively desaturated air, and the main bulk of the evaporator water is cooled by this evaporation. A heat-exchanger coil is introduced to utilize this cooling action.

(2) A generator is introduced into the system to maintain the salt solution at a proper concentration. This is accomplished by pumping the weak solution from the absorber to the generator, where the excess water is boiled off. (This boiling action accounts for the heat requirement that was noted above. This heat source may be steam generated by the winter heating system, or it may be high intensity waste or exhaust heat from some other system.)

The re-established strong salt solution is then pumped back to the absorber to repeat the absorption action.

(3) A condenser is utilized to process the water vapor that is boiled off in the generator. The vapor condenses back to liquid form (water) as heat is rejected from the condenser to the heat sink or cooling tower. This water is then returned to the evaporator, where the process begins again.

Throughout this cycle, the only moving parts are the circulating pumps.

FIGURE 2–3.20

THE ABSORPTION REFRIGERATION CYCLE

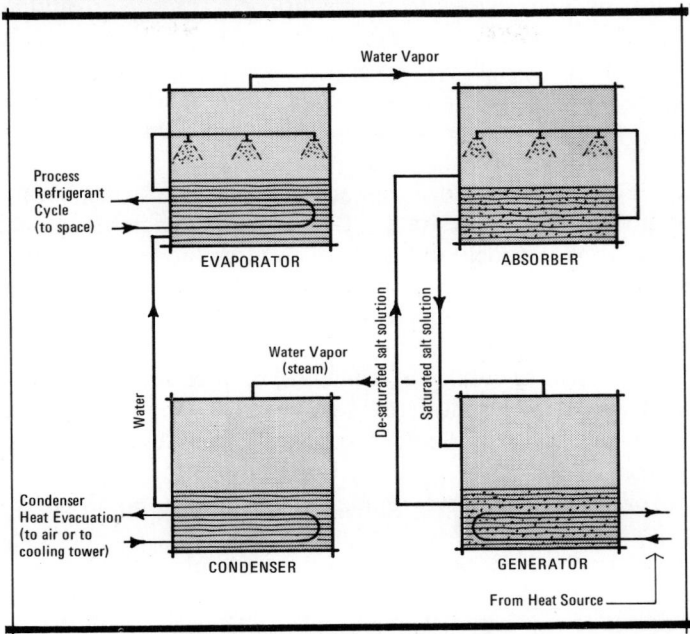

Fan-coil devices The cooling medium within the conditioned space itself is generally air (although water is sometimes used for panel cooling systems). But the processing of this air involves a route of process refrigerant flow that is generally separate from the internal refrigeration cycles just described.

The point of contact between these two cycles is a heat exchange coil that passes through the evaporator (see Figure 2–3.19). A process refrigerant passing through this coil will be cooled by the action of the evaporator; and heat taken from this process refrigerant provides the *heat of evaporation* required in the basic refrigeration cycle.

The resulting chilled process refrigerant is then pumped to a point in or near the space to be conditioned. At this point, a fan circulates warm room air (return air), outdoor air, or a mixture over a coil which concentrates the chilling potential of that refrigerant. This action extracts heat from the warm air, thereby cooling and dehumidifying this air and preparing it for circulation into the conditioned space.

The extracted heat that is absorbed by the process refrigerant then moves back to the evaporator area, where the cycle begins anew.

Humidity control. Temperature reduction and dehumidification are both accomplished by the same evaporator coil (see fan-coil in Figure 2–3.19). In principle, the air to be conditioned (example: outdoor air at 80°F, 70% RH) is drawn over the cooling coils. This action cools the air to a near saturated condition. Excessive water vapor is condensed and drained away; and the air may then be mixed or reheated as necessary to produce an optimum environmental balance (typically 75°F, 50% RH).

Filters. For control of localized smoke, dust, or fumes, local exhaust vents and hoods may be desirable. But in a large number of cases, the source of pollution is widely diffused (or even unknown). For this reason, filters are typically required within the air handling system to process both recirculated air and outdoor intake air.

Such filters may take several forms:

(1) *Dry filters* such as cloth, felt, cellulose, or wire screens are a common type. These elements must be kept clean, so a regular maintenance program is required. Some types are disposable to facilitate this process.

(2) *Spray washers* also serve to regulate humidity. So these two functions can be combined to some extent.

(3) *Electric precipitators* are based on the principle of attraction of static charges. They are effective for very fine dust, and may involve an automatic system to wash the collecting plates.

Thermoelectric devices for cooling The thermoelectric principle requires no refrigerant fluid. Nor does it involve mechanical action within the refrigeration device itself.

When an electric current is passed through a thermocouple of two dissimilar materials, the junction temperatures will vary. Heat is released at one junction, while heat is absorbed at the opposite junction (see Figure 2–3.21). If the current is reversed, the heating and cooling effects are also reversed. This phenomenon is called the *Peltier Effect,* and this forms the basis for thermoelectric cooling.

Insulators are poor thermoelectric materials because their electrical conductivity is small. Metals also fail because their thermal conductivity is high, while their thermoelectric power is low. The best results are obtained with semiconductor materials that exhibit properties midway between those of metals and insulators.

With the proper choice of materials, then, the Peltier Effect can be utilized for heating and cooling purposes. And this system is one that performs in immediate response to electric current, without the need for an intermediate refrigerant vehicle.

FIGURE 2–3.21

THE THERMO-ELECTRIC CYCLE

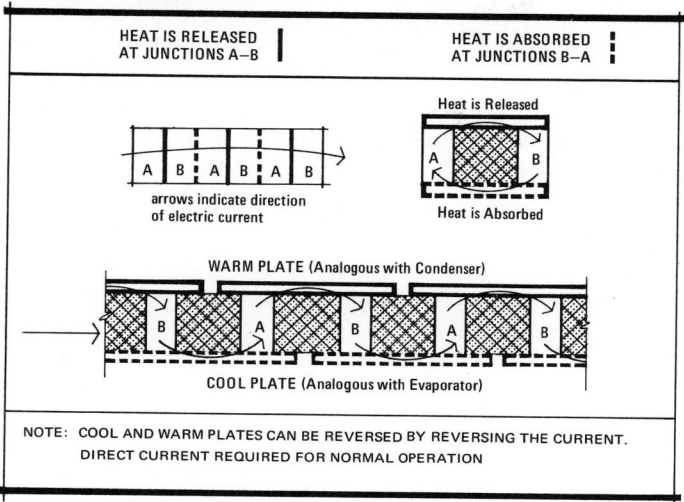

NOTE: COOL AND WARM PLATES CAN BE REVERSED BY REVERSING THE CURRENT.
DIRECT CURRENT REQUIRED FOR NORMAL OPERATION

Demand zones

Because of the variable nature of solar loads and internal loads, interior heating and cooling demands may be quite variable at a given point in time. *Demand zoning* is a method that is used to compensate for this load disparity.

Generally, the development of *demand zones* will depend on the nature of loads that impinge on the enclosed space. Some loads exert a somewhat constant influence, in the sense that all parts of the building interior are affected in a similar way. Conduction losses through wall and roof assemblies, and distributed internal heat gains such as electric lighting are two examples of thermal loads that tend to be somewhat consistent throughout the building. If these are the only significant loads affecting internal comfort (as they conceivably could be for buildings that have no window openings), then the lower floors of the building could be developed as a single demand zone for control purposes (excluding the influence of the roof). Minimum thermostatic controls and dampers would be required to maintain comfort because the system can be easily balanced and demand will not fluctuate significantly within the enclosure.

But when wall and roof openings are prevalent in the design, solar and wind loads may vary considerably in different portions of the building. For example, the effect of solar heat gains vary as the south, east, and west walls are alternately exposed to the direct rays of the sun. Solar loads also produce seasonal variations, such as the fact that south rooms may experience a heat gain on a clear day during the winter months, while adjacent north rooms are experiencing a simultaneous heat loss. Comfort demands will require that the heating and cooling systems be sufficiently flexible and controllable to respond to such simultaneous demands.

Ideally, for most common building types, the optimum control procedure would be to make each room or definable area an individual demand zone; with a thermostat to control system performance in each of these areas. But the matter of economics must be considered with this *ideal,* and many building programs cannot support such extensive control costs. It then becomes a problem of subdividing the building, to group rooms or areas that are consistently exposed to similar thermal influences.

FIGURE 2–3.22

VERTICAL AND HORIZONTAL DEMAND ZONING

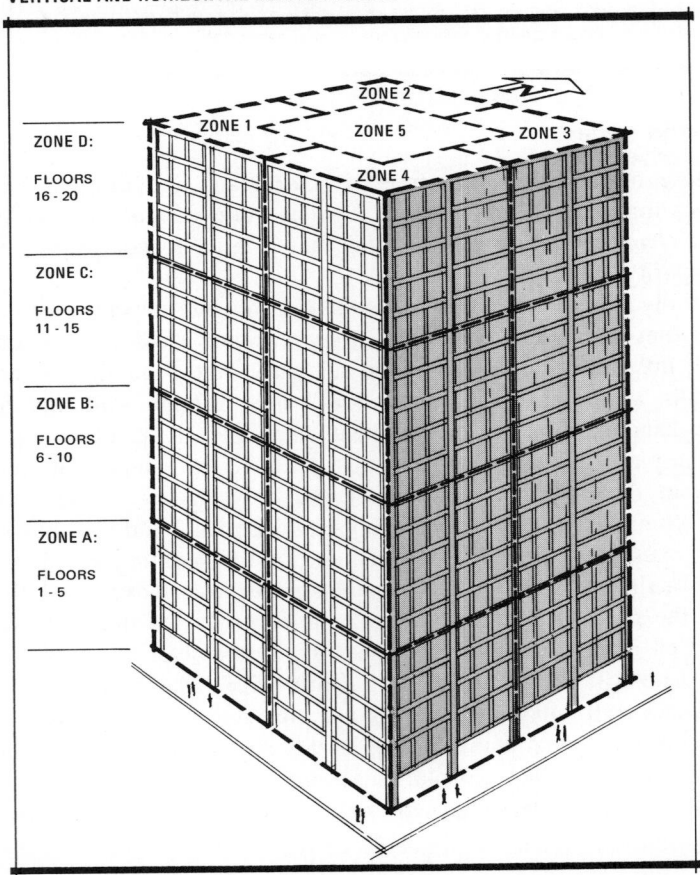

To do this, areas that are exposed to variable thermal conditions should be set apart (mechanically) from those that are not, and from those that are exposed to different thermal variables. In this way, demand zones will tend to define themselves, with all of the rooms in each zone experiencing similar thermal influences.

This procedure may involve a horizontal subdivision—such as east, west, south, and north exposures, plus interior areas that have no external exposure (see Figure 2–3.22). Zoning can also involve a vertical subdivision—to compensate for vertical variations in activity or occupancy, to produce more optimum utilization of mechanical equipment, or to overcome the *stack effect* in tall buildings.

Distribution systems

The distribution systems within an enclosure must therefore respond to changing thermal demands in an attempt to regulate the environmental characteristics of the interior space (as generally discussed in Chapter 1–3).

To accomplish this, a heat transfer cycle must be developed that (1) enables the internal air mass to be periodically and methodically moved past the critical room surfaces(where heat transfer takes place), (2) then to be exhausted from the immediate occupied area by moving the air flow to a more remote mechanical room location where it can be processed, and finally (3) to be reintroduced back to the occupied area where the cycle is repeated.

Thermal potential in the demand zone The potential of supply air for modifying the temperature of an interior air mass is dependent on two factors: (1) the volume-rate of the supply and return air, and (2) the difference between the temperature of the supply air and the temperature of the air mass in the space that is being conditioned. The following formula is useful for estimating the combined influence of these variables:

$$H_a = 0.018V (T_2 - T_1)$$

where: H_a = the heat transfer capability of the supply air (btu/hour.)
V = the volume of supply air (cubic feet/hour.)
T_1 = temperature of the room air mass (°F)
T_2 = temperature of the supply air (°F)

The manipulation of one or both of these two variables is the basic function of the air-handling system.

The Constant Volume Dual-Duct System. The constant volume dual-duct distribution system is a specific response to the zoning requirements of temperate zone building that have moderate-to-extensive glass areas. These buildings must respond to extremely diverse climatic and solar conditions. For example, as noted previously, a significant winter heat loss on the north wall can occur concurrently with a solar heat gain on the south wall. This condition can change quickly; for when a cloud obscures the sun, the south wall will quickly shift from heat gain to heat loss conditions similar to the north wall. Similarly, hourly demands shift significantly

FIGURE 2–3.23

THE DUAL DUCT SYSTEM

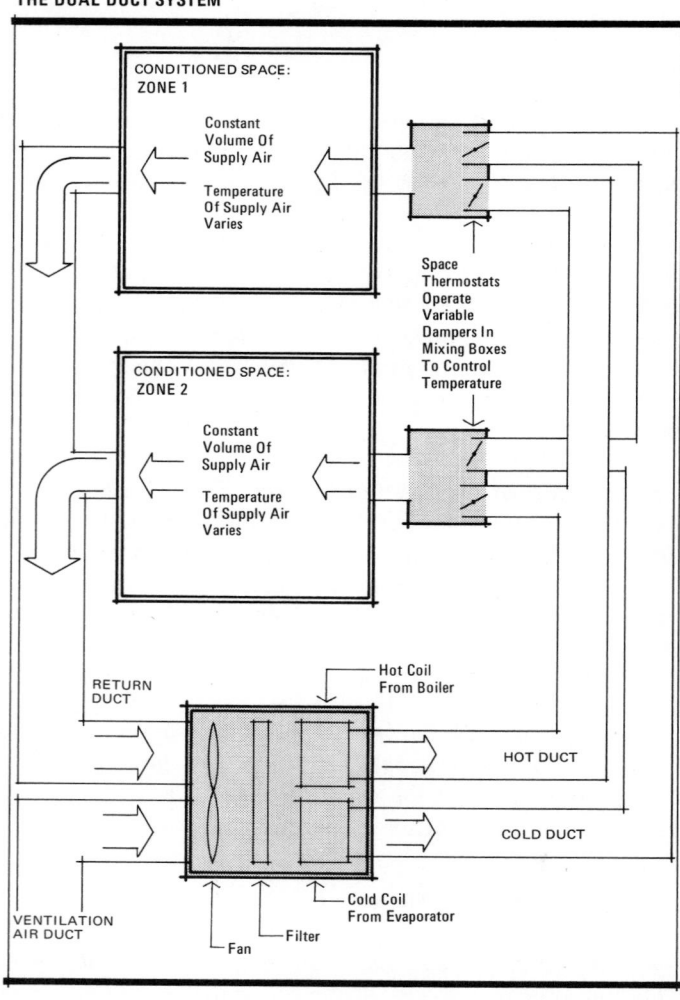

during the day as the sun position changes from the east face to the west face of the building.

The constant volume dual-duct system responds to these changes by providing simultaneous parallel supply air flows of hot and cold conditioned air (see Figure 2–3.23). Periodic variations in room demand conditions are met by varying the mix; and this is accomplished at thermostatically-operated *mixing boxes,* where hot and cold air is mixed to provide the proper supply air temperature before it enters the space to be conditioned.

This concept provides a constant volume of air to the space. The thermostatically-controlled variable, then, is supply air temperature.

While this concept is effective in providing quick response to varying demand conditions that occur simultaneously within a building, there may be an inherent inefficiency in the need to provide *both* hot and cold air continually to meet the extreme variations that commonly occur during the cold and intermediate seasons. The process of reheating previously cooled air, or the inverse process of cooling previously heated air, can introduce inherently inefficient procedures.

The Variable Air Volume System. The less costly, but less flexible concept of the variable air volume system involves a single supply duct, with processed air circulated at a constant temperature (see Figure 2–3.24).

The air mass in the conditioned space is therefore subjected to a variable rate of air change, and this must vary within prescribed limits. The air change rate must be high enough to avoid stagnation and contamination (see Table 1–3.6), and yet, not so high as to introduce drafts and cold spots. This latter limit is generally near 20 air-changes-per-hour for high velocity supply systems (a limit that also applies for constant volume systems).

A major limitation of this system is that response to simultaneous heating and cooling demands cannot be provided—as it can with the dual-duct systems. For this reason, the system is generally incapable of providing a completely adequate response to the periodic external changes that call for varying heating and cooling demands in a building—particularly those that occur during the intermediate seasons. For these reasons, the use of this system tends to be limited to interior zones, and to building designs that minimize the influence of changes in direct solar radiation.

Panel cooling for zoning control (with variable air volume systems). It has been estimated that a ceiling surface maintained approximately 15° cooler than the room air and other surfaces in the room will absorb sen-

FIGURE 2–3.24

THE VARIABLE AIR VOLUME SYSTEM

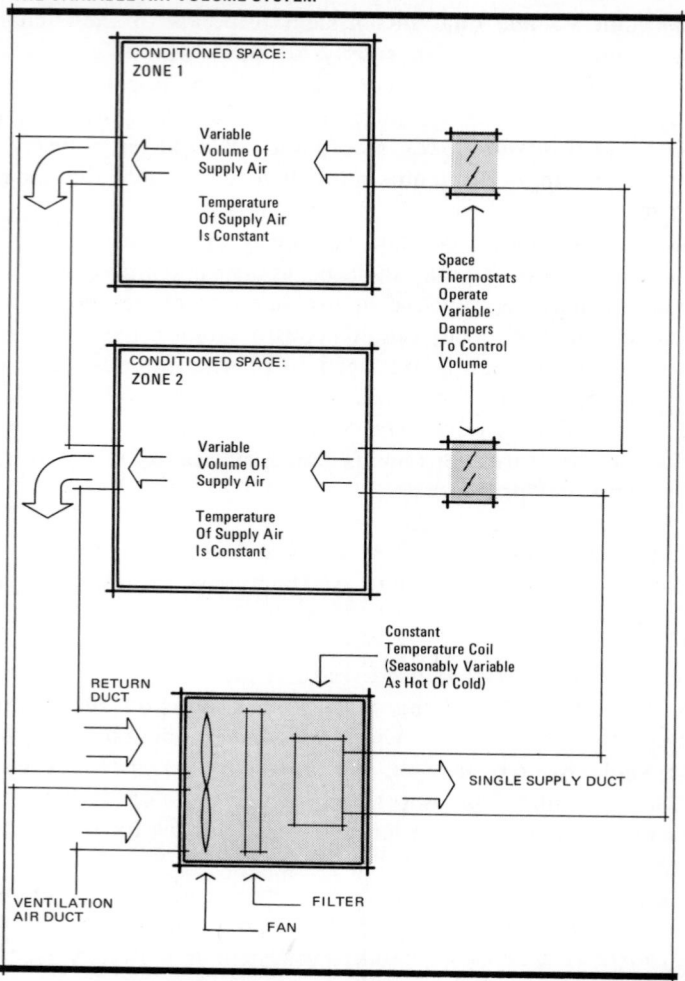

sible heat at a rate of at least 20 btu/sq ft/hr. When window areas are well-shaded or constitute limited portions of the facade, this heat control capability can approximate the differential between typical interior and perimeter zone demands.

This principle can diminish the previously discussed limitations of the variable air volume system. If a zoned perimeter *panel cooling* system is used in conjunction with this air distribution system, it is possible to maintain consistent thermal control with a single system of fans and ducts (probably supplemented by small convectors under the windows for winter use only).

With somewhat constant air handling requirements, then, demand zone control can be achieved entirely by means of: (1) chilled water supplied to the radiant ceiling panel in warm weather, or (2) heated water supplied to the perimeter convectors in winter.

Three- and Four-Pipe Water Systems. The 3- and 4-pipe water systems function in a manner that is similar in principle to the dual-duct system. But where the dual-duct system utilizes hot and cold *air* as the distribution medium, the 3- and 4-pipe systems utilize parallel hot and cold *water* as the distribution vehicle (see Figure 2–3.25).

FIGURE 2–3.25

THE THREE-PIPE WATER SYSTEM (WITH INDIVIDUAL FAN COILS)

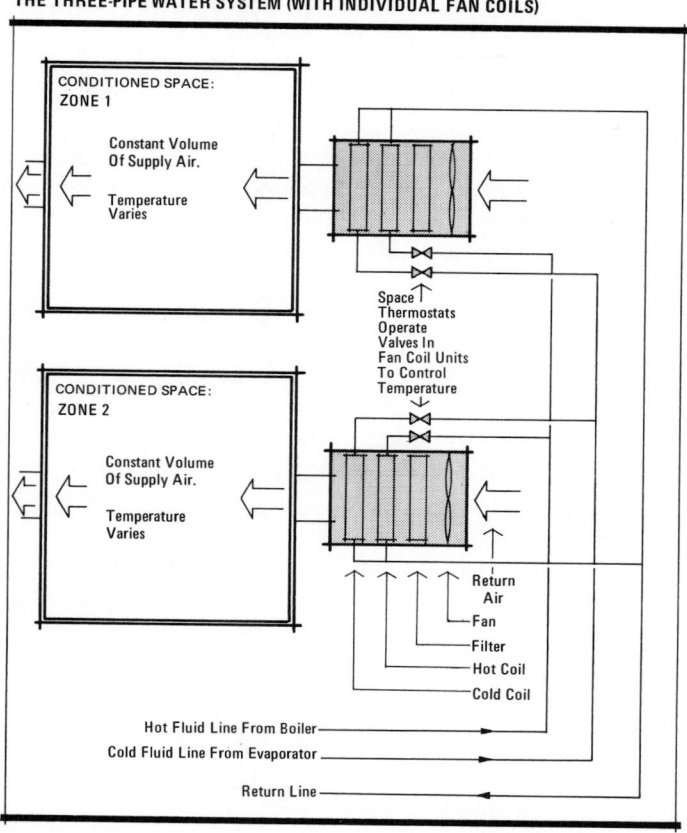

Since water is much more compact and efficient as a medium for heat transfer, the space savings are often very significant. Large supply and return air ducts are replaced by much smaller piping. (The use of a single return pipe or separate hot and cold return pipes constitutes the distinction between 3- and 4-pipe systems.)

Like the dual-duct system, these water systems are effective in providing quick response to the variable demand conditions that occur simultaneously in buildings that have moderate to extensive glass areas.

Distribution of air within the occupied space The ultimate objective in providing refrigeration or heating capability is to influence occupant comfort, and this decisive thermal action must obviously take place within the occupied space itself.

Within the occupied space, then, the system must be capable of adding or extracting heat at critical surfaces and at the immediate occupant location. This is generally done by processing the entire interior air mass; and this air mass becomes the major vehicle for environmental manipulation.

Natural Convection Due to Gravity. Localized volumes of stagnant air will cause erratic temperature and humidity conditions within the interior air mass. Similarly, concentrations of heat gain or heat loss can cause localized discomfort in a space, in spite of the fact that the average room conditions are well within the comfort zone. For both of these reasons, then, it is necessary to study air paths in the space carefully in order to insure that air passes through or induces proper movement in each portion of the space.

Generally, this involves a recognition of the natural (gravity) tendency of cool air to seek a lower level and warm air to rise. Figure 2–3.26 indicates the significance of this natural air convection in the diffusion or concentration of cool air in the immediate environment of the occupant. Such convective influences are controlled, in part, by the placement of air supply and return elements; in this case, by the placement of the heating element.

Forced Convection. The projection and directional capabilities of forced air systems can be utilized to modify and extend the simple flow patterns induced by natural gravity convection.

In this regard, the supply air is considered to have two basic characteristics that determine its performance: (1) motive force, and (2) thermal character (primarily temperature). At present, air duct velocities of up to 6000 ft/min can be handled efficiently and with relative quiet. Supply air temperatures of up to 160°F can be used in cold weather; and

FIGURE 2–3.26

NATURAL GRAVITY CONVECTION

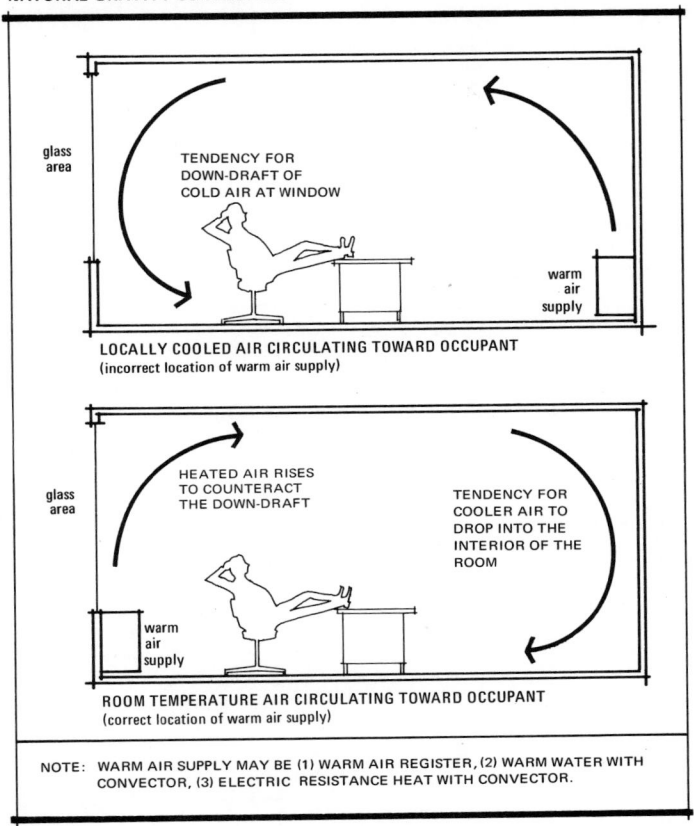

LOCALLY COOLED AIR CIRCULATING TOWARD OCCUPANT
(incorrect location of warm air supply)

ROOM TEMPERATURE AIR CIRCULATING TOWARD OCCUPANT
(correct location of warm air supply)

NOTE: WARM AIR SUPPLY MAY BE (1) WARM AIR REGISTER, (2) WARM WATER WITH CONVECTOR, (3) ELECTRIC RESISTANCE HEAT WITH CONVECTOR.

while warm weather supply air tends to be limited by humidity factors (condensation on ducts, etc.), supply temperatures 20–30° below room air temperature are not uncommon. In principle, the greater the air velocity and the greater the temperature difference between supply air and room air, the greater will be the potential reduction in duct sizes.

But care must be taken to avoid high velocity air streams within the occupied zone in the room—this zone being generally defined as the entire room below the 6-foot level. This limitation tends to lead to the use of ceiling or high wall supply outlets (see Figure 2–3.27).

As shown in the figure, discharge flow may be quite low in velocity. Or, where high velocity discharge is involved, this air should be introduced in a way that will develop a horizontal or near-horizontal flow across the

FIGURE 2–3.27

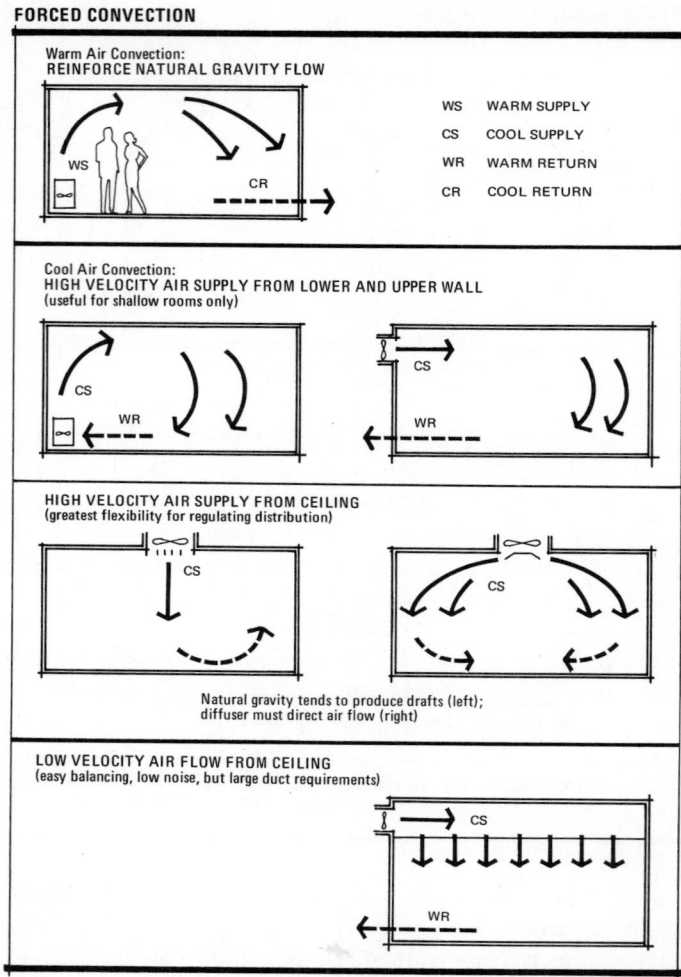

FORCED CONVECTION

Warm Air Convection:
REINFORCE NATURAL GRAVITY FLOW

WS

CR

WS   WARM SUPPLY

CS   COOL SUPPLY

WR   WARM RETURN

CR   COOL RETURN

Cool Air Convection:
HIGH VELOCITY AIR SUPPLY FROM LOWER AND UPPER WALL
(useful for shallow rooms only)

CS

WR

CS

WR

HIGH VELOCITY AIR SUPPLY FROM CEILING
(greatest flexibility for regulating distribution)

CS

CS

Natural gravity tends to produce drafts (left);
diffuser must direct air flow (right)

LOW VELOCITY AIR FLOW FROM CEILING
(easy balancing, low noise, but large duct requirements)

CS

WR

ceiling. In this way, the supply air is permitted to diffuse and interact with the upper level strata before it moves down to the occupied zone at reduced velocity and with a reduced temperature gradient.

Concerning the performance of air after it leaves the diffuser, it should be noted that a typical air jet that is blown horizontally into an open area (with no temperature difference between supply air and room air) will tend to expand approximately 15° on either side of the axis of travel.

When supply air temperature is below the temperature of the air in the room (i.e., a cooling influence), density differences will deflect the trajectory downward due to the effects of gravity. Conversely, the introduction of warmer supply air will tend to deflect the trajectory upward. In either case, the *throw* or horizontal projection distance will vary with the discharge velocity of the air jet. Velocity may also affect the angular spread of the cone of air described above. But in general terms, this combination of phenomena can be used to estimate the approximate pattern of air movement from a given diffuser location.

Condensation Control. Air paths also exert an evaporating action that is important in controlling condensation.

This is most apparent in the need to ventilate concealed structural spaces (such as wall, ceiling, and floor cavities). This ventilation is needed to facilitate the removal of water vapor that may pass from occupied areas into these concealed locations. (See previous discussion of vapor barriers in this chapter.) Without adequate ventilation, vapor that condenses in the colder concealed space can cause water damage, and in some cases, damage due to frost or freeze conditions.

In a related sense, warm air flow up the inside face of a window tends to warm the glass, thus somewhat reducing the tendency toward condensation in cold weather. In order for this flow of air to be most effective, however, it must be constant, not intermittent air as may be supplied to the main space in response to thermostatic control. Constant output supply systems may therefore be required in these perimeter locations.

Limitations affecting air processing within the occupied space

As more heat is introduced into the occupied space, this heat can be neutralized by increasing the rate of air change. Typically, high velocity supply air can be introduced and increased in volume up to approximately 20 air-changes-per-hour. When this volume limit is approached and exceeded, there may be drafts and cold spots in the occupied zone.

The other alternative for neutralizing internal heat gain is to introduce colder air. But this approach also has a limitation. As the input air is cooled below the dew point in the room, moisture will begin to condense on ducts and metal diffuser surfaces. This condition establishes a limit of approximately 20° as the maximum temperature difference between the cooled input air and the air in the room.

When the designer increases the glass area or otherwise diminishes the ability of the building shell to intercept external heat, or when he increases internal loads such as those associated with computers or electric lighting, the thermal load inside of the building increases. At the same time, the mechanical engineer is confronted with limitations in the quantities of heat that can be neutralized within the occupied zone by conventional air processing methods. This limit generally falls in the range of 60–75 btu/sq ft of floor area.

Supplemental panel cooling techniques offer one possibility for extending this limit. But in general, it becomes advisable and economically necessary to evaluate techniques for isolating and removing waste heat more effectively and efficiently; to harness waste heat and minimize the impingement of this excess load within the occupied space itself.

Minimizing Heat in the Occupied Zone. In some cases, substantial quantities of waste heat can be locally isolated. Heat can be trapped within the lighting system, absorbed by a solar screen, isolated in the immediate vicinity of heavy equipment or computers, etc. When this principle is applied, it develops that significant quantities of heat are not immediately incident in the occupied part of the room at all. Rather, this heat is initially confined in an enclosure, often in a ceiling or wall cavity that is somewhat remote from the part of the room where people are normally located. This heat is in a form and is in a location where it can be isolated and controlled with air or water. (Note that this is a variation from the more conventional method of providing cooling capacity to neutralize heat gains *after* this heat becomes a direct influence within the occupied space itself.)

The precise economic value of the isolation principle will depend on the concentration of energy involved (watts/sq ft; area of glass; etc.). But from the standpoint of the total sensory environment, the principle is significant because it permits the introduction of heat-producing elements (electric and natural) into the interior space with a minimum of uncontrolled heat in the occupied zone. This effectively extends the internal heat control potential beyond the previously noted 60–75 btu/sq ft limit. It also means that internal heat gains can be harnessed; to be exhausted or recirculated and utilized, depending on the nature of external influences on the building shell.

The Influence of Controlled Lighting Heat on Interior Cooling Performance. The thermal performance of a general lighting system will demonstrate the potential of this heat isolation principle.

With continuous operation, all of the lamp energy is eventually imposed as heat on the occupied space, unless an attempt is made to isolate and control it. Figures 2–3.28A and 2–3.28B compare conventional and luminaire exhaust systems for a representative fluorescent lighting system of 6 watts/sq ft.

The basic operation of the luminaire exhaust system is as follows:

(1) Room air is drawn (by negative pressure) through the luminaire face, passed over the lamps, and exhausted into a return duct or plenum above. With this action, a substantial proportion of the lighting heat is removed before it can enter the occupied space. (See previous discussion, "control of electric lighting heat" in this chapter.)

(2) The warmed return air can then be handled in one of several ways, depending on the thermal demands of the building at a given time:

- Warm return air can be totally or partially rejected to the outside in moderate weather when outdoor temperature and humidity conditions provide an unlimited supply of high quality replacement air.
- Return air can be retained in the building; to be tempered and regulated through proportional mixing with cooler outdoor air or with refrigerated air as required. In some cases, this warm return air can substitute for the warm air portion of a dual-duct system.
- In hot-humid conditions, the entire quantity of return air can be recycled through the refrigeration system before it is redistributed.

The Influence of Controlled Lighting Heat on Interior Heating Performance. Climate and building design parameters will determine the relative influence of lighting heat in cold weather. But in general, this heat is useful provided it can be readily redistributed as needed. (See Figure 2–3.29.)

The basic operation of the luminaire exhaust system on the heating cycle is as follows:

(1) Room air is drawn (by negative pressure) through the fixtures, passed over the lamps, and exhausted into a return duct or plenum above.

(2) The warmed return air can then be handled in one of several ways, depending on the thermal requirements of various rooms at a given time:

- Return air can be redistributed to occupied areas and released as warm air through air supply diffusers.
- Return air temperature can be tempered and regulated through proportional mixing with cooler outdoor air.
- Return air heat can be extracted by passing it through a heat exchanger (the extracted heat to be used for tempering outdoor ventilation air, etc.).
- Although low levels of transfer are involved, return air heat can conceivably be transferred to a water circuit, and the heated water can be stored for use during periods of higher heat demand (for example at night, when lighting is off).

FIGURE 2–3.28A

COOLING CYCLE:
CONVENTIONAL AIR RETURN

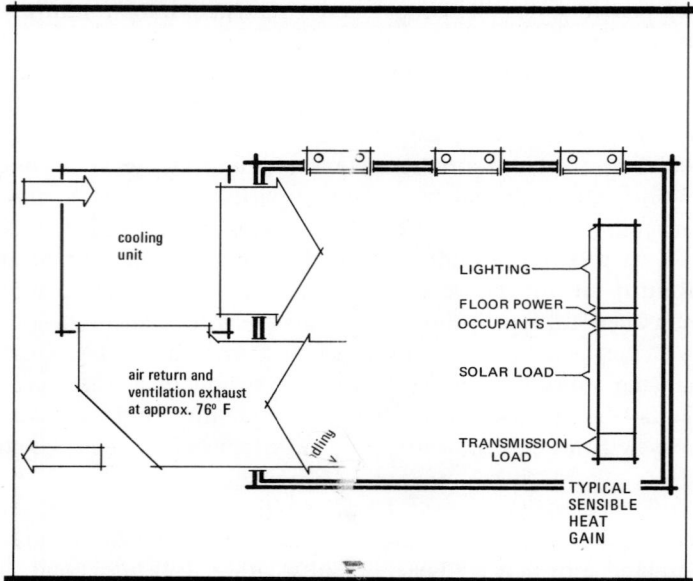

FIGURE 2–3.28B

HEATING CYCLE:
AIR RETURN THROUGH LIGHTING SYSTEM

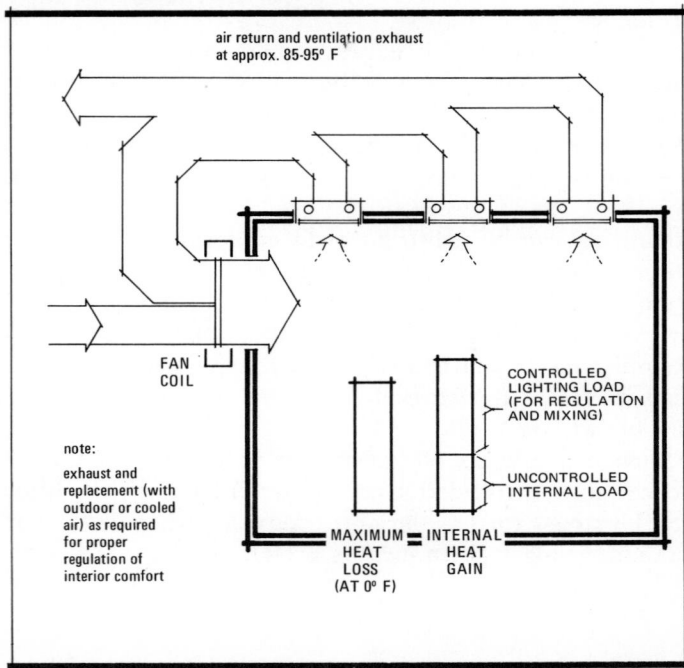

FIGURE 2–3.29

COOLING CYCLE:
AIR RETURN THROUGH LIGHTING SYSTEM

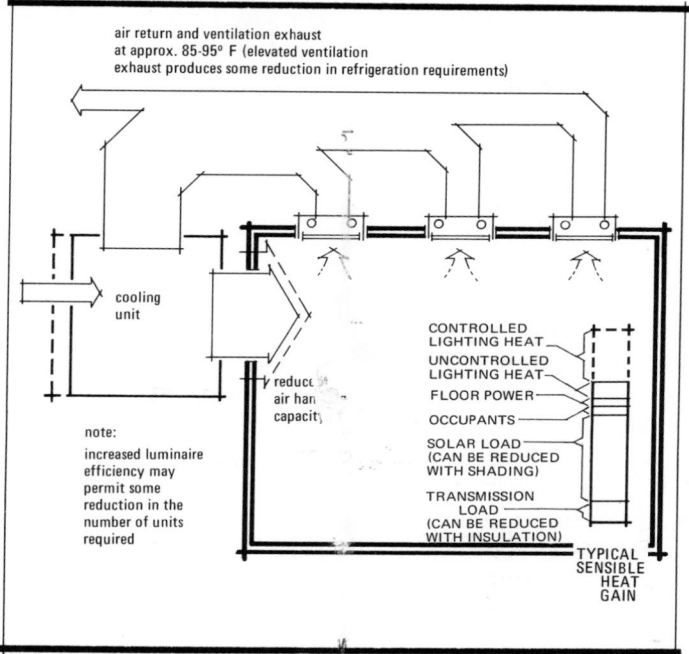

Radiant energy systems The immediate thermal environment can also be adjusted by manipulation of *mean radiant temperature* (see Chapter 1–3). In this sense, there are two principal techniques: (1) panel radiation, and (2) concentrated radiation.

Panel Systems. Ceiling and floor surfaces can be utilized as supplementary or primary panel radiation devices. Walls can also be utilized, but are generally of limited value because of the varying and inconsistent distances from a wall panel to the occupant, and because of the variable area available for use in this way (being limited by window openings, door openings, shelf and closet locations, etc.).

Probably more than any other form of thermal energy distribution, panel systems must conform to the specific characteristics of the interior-exterior thermal exchange. Pipe coils should be spaced closer together near window areas, with wider spacing in the more interior areas. Furthermore, effective insulation should be positioned to minimize the inefficiencies associated with losses to the ground, to the exterior air, or to unoccupied spaces.

This method is limited by the maximum values for floor and ceiling temperatures described in Chapter 1–3. As a general rule of thumb, the capacity of a floor panel should be limited to approximately 10–20 btu/ sq ft of surface. For ceilings, the heat capacity limit is approximately 70 btu/sq ft of surface. When the room heat loss exceeds this limitation, other means of primary or supplementary energy distribution must be utilized.

The principal disadvantage of this technique is the fact that control is limited to temperature only. Such factors as humidity and atmospheric contaminants are not affected by this process.

Concentrated Thermal Radiation. For areas or situations where the frequency of air change precludes an attempt to process the environmental air mass itself, high intensity sources of infrared energy can be effectively controlled and concentrated for limited-zone heating. Typical areas are garages, uninsulated warehouses, semi-protected waiting shelters, and building entrances.

Essentially, these systems involve compact high intensity energy sources that can be placed at the focal point of a specular parabolic reflector. In this way, narrow or broad directional cones of energy are emitted. To be effective, these systems should be designed to maximize surface exposure within the beam (see Figure 2–3.30).

FIGURE 2–3.30

CONCENTRATED RADIANT HEAT

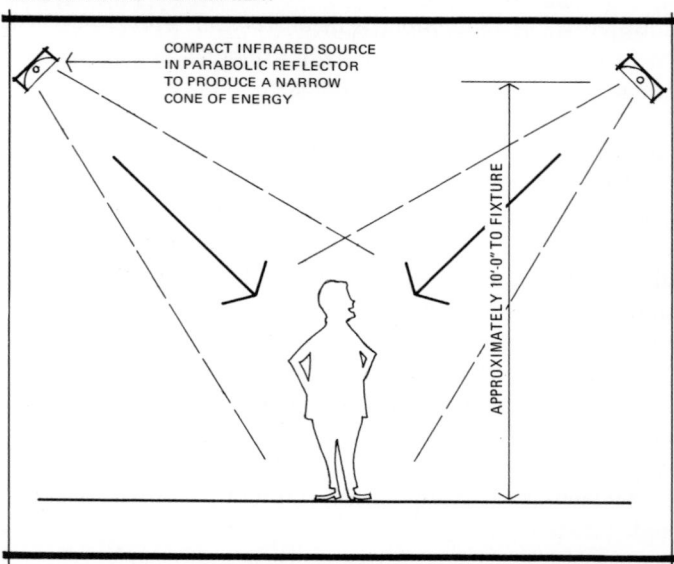

COMPACT INFRARED SOURCE
IN PARABOLIC REFLECTOR
TO PRODUCE A NARROW
CONE OF ENERGY

APPROXIMATELY 10'-0" TO FIXTURE

This technique is similar in principle to the warming effect of the sun on calm days in winter when the air temperature is relatively low. These concentrating energy sources, then, can be used to heat semi-exposed and localized areas independent of the environmental air temperatures involved, provided the object or person to be heated is effectively screened from wind. These sources can also be used in situations where quick *warm-up* response is required.

Table 2-3.19 Representative Limited-Zone Radiation for Interiors

Temperature of Indoor Air	Supplementary Incident watts/sq ft (Required to Compensate for the Deficient Indoor Air Temperatures)
60°F	10–25 watts/sq ft
50°F	20–35 watts/sq ft
40°F	40–55 watts/sq ft

NOTE: Lower intensity in above ranges apply for nondraft interior areas, where occupants are separated from cold walls.
Higher intensity in above ranges apply for drafty interior areas, where occupants are exposed to cold wall conditions nearby.

NOTE: Wattages shown apply for normally-clothed interior occupants (when outdoor air temperature is 0°).

Reduction of Internal Radiant Heat Effects. There is often a radiant heat effect associated with the electric light beam in a space, particularly when incandescent systems are involved.

During the cooling season, this radiant energy is difficult to eliminate with the usual air circulation methods because the transfer of heat is directly between two objects or surfaces and does not directly involve room air. In this sense, the following techniques are useful for reducing such radiant heat effects:

(1) *system efficiency.* Utilize more efficient devices, or those that emit less infrared energy. For example, substitute fluorescent for incandescent units.

(2) *surface temperature.* Reduce temperature gradients between surfaces, objects, and occupants. For example, use panel cooling or controlled air circulation to cool the metal surfaces of an offending device or fixture.

(3) *air temperature.* Maintain the room air temperature slightly cooler than normal, so the gains and losses in body heat are in better equilibrium. (See Chapter 1–3 for discussion of "mean radiant temperature.")

(4) *selective control of energy.* Divert invisible infrared energy from the occupied zone by selective absorption, reflection, or transmission. For

example, use *interference filters* that transmit light and reflect the invisible radiation. Infrared in the beam can be reduced as much as 85% by this technique with no significant loss of light. (However, for interior devices, the redirected heat must be deflected to a heat-absorbing collecting surface, which in turn must be cooled by more conventional air or water techniques.)

PART 3

Coordinated System Development

Building design continues as a challenge for improvement and excellence. At present, there is concern with the problem of consolidating and coordinating the significant technological achievements of the past century; to incorporate them into our vocabulary of design and form.

Much of the substance of the currently identified design vocabulary can be traced to the ideas and example of the Bauhaus (Germany) in the 1920s. This group taught designers to recognize the machine as a rising influence in western culture and building. Mass production, component assembly, functional simplicity, and modular attitude were shown to have logical and useful design implications.

But during the several decades since such theories became sufficiently clarified to be recognized and accepted in our society, increasingly sophisticated technical developments have been introduced and perfected. Questions are raised concerning the significance of these and the changes they may portend in our approach to building design.

Some of these questions evolve from the fact that the architect today has greater freedom than ever before in reference to manipulation and control of sensory perception and comfort. But this freedom cannot be accepted as license for undisciplined design. In this sense, the search for a sense of direction must begin by eliminating the question, "How do you light and air condition a preconceived building design?"—and replacing it with the question, "How do you design a building that is in sympathy with the contemporary capability to manipulate and control the interior sensory environment?" The building form that results from such a question should relate to both natural and artificially-generated environmental influences; and Part 3 will attempt to summarize and define some of the influencing factors.

The Building as a Comprehensive Environmental System

Throughout history, one of the limiting criteria in building design has been the need to bring light and air into an enclosed space, and to protect the interior from adverse external influences. Each society and culture has produced its own solution to this problem, reflecting climatic demands as well as the technology and ingenuity of the time.

As in the past, the contemporary architect must also resolve this problem of environmental performance. For example, he can develop his enclosure to utilize natural light, natural ventilation, and beneficial solar influences during under-heated periods of the year. In this sense, he is continuing the long sequence of experiments and developments that have evolved throughout the history of building.

But several factors make this period an obvious break with the past. This century has been a continuing evolution of products and mechanical techniques for environmental control—electric lighting, mechanical air distribution systems, heating and cooling devices, and sound control. Viewed as parts of a comprehensive system, these techniques combine to provide a revolutionary new freedom to regulate the interior sensory environment. With this combination of tools, then, the architect finds that the design of the exterior building shell is no longer *necessarily* subject to the uncompromising limitations of natural ventilation, daylighting, and solar control. And the occupants of the building find that their comfort and activities are increasingly independent of natural outdoor conditions related to climate, weather, or time of day.

So architecture reflects, in part, man's continuing attempt to establish a protected environment that approximates the conditions in which he is most comfortable and at ease; in spite of the fact that such comfort conditions appear only intermittently in nature.

But rather than a simple correction of climatic deficiencies, the environmental control function of the building must be oriented toward the more extensive sensory demands of various occupant activities and experiences.

This occupant perceives light as surface brightness and color; he absorbs heat from warmer surfaces and warmer air; and he himself emits heat to cooler surfaces and cooler air. He responds physiologically to humidity, to air motion, to radiation and air *freshness*. He also responds to sound. A major function of the building, then, is to provide for all of these sensory responses concurrently—to establish and maintain order and harmony in the sensory environment.

THE DIAGRAMMATIC PLAN AS AN ORGANIZATIONAL TOOL

The *diagrammatic* plan is a useful planning tool for identifying, defining, and communicating among consultants the multiple and interdisciplinary demands involved in analysis of the man-made environment. In sequence, this design step comes after the basic building program is known (activity needs, etc.). It precedes the development of *schematic plans* (plans that begin to take definite, though preliminary, form and dimension).

In this sense, the diagrammatic plan indicates basic relationships within and between various activities, without implying either form or dimension. It is used to generate a definition of the building in terms of general performance requirements, and thus precedes the generation of a specific building design.

In part, the diagrammatic plan serves as a relationship diagram which explores (without implying form) layout alternatives to be considered in subsequent planning stages. This tool also begins to identify the relative location of potential barriers and transitions.

Figure 3–1.1 indicates a representative diagrammatic plan for a grouping of five activities. (These activities are simply identified as numbers in this discussion.) Such studies will facilitate a preliminary performance analysis of each separate activity (1,2,3,4,5). They will also facilitate a study of barrier or transition relationships associated with each of these activities (1–2, 1–3, 2–3, 2–4, etc.).

Figure 3–1.2 takes this a step further and begins to explore the nature of the service distribution network.

The Comprehensive Performance Specification

Within the context of this book, the diagrammatic plan is a particularly useful organization tool for defining preliminary performance specifications to guide the ensuing system development. Continuing the example illustrated in Figure 3–1.1, the pattern diagrams being studied here provide a framework for defining and evaluating the factors noted in Tables 3–1.1 through 3–1.4.

FIGURE 3–1.1

THE DIAGRAMMATIC PLAN: ACTIVITY RELATIONSHIPS

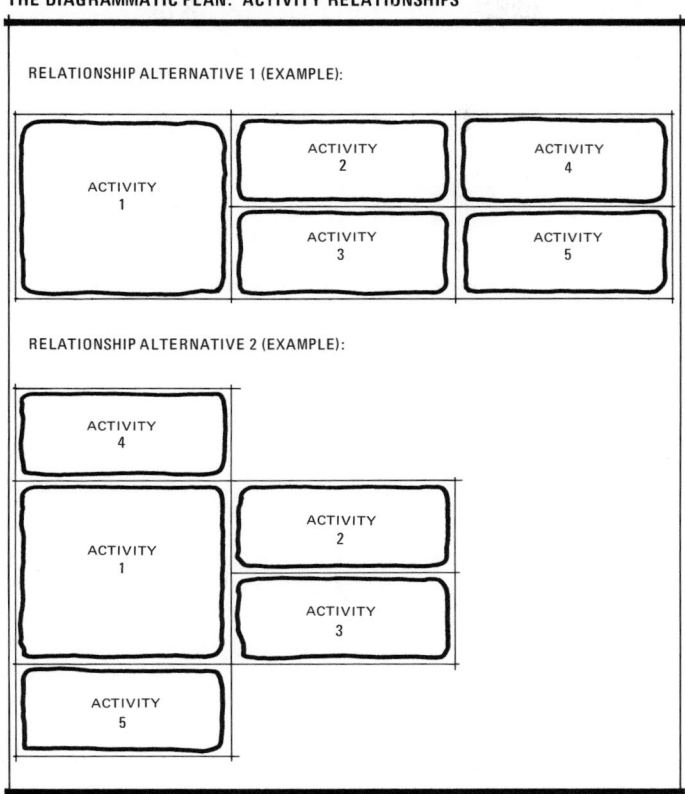

RELATIONSHIP ALTERNATIVE 1 (EXAMPLE):

ACTIVITY 1

ACTIVITY 2

ACTIVITY 3

ACTIVITY 4

ACTIVITY 5

RELATIONSHIP ALTERNATIVE 2 (EXAMPLE):

ACTIVITY 4

ACTIVITY 1

ACTIVITY 2

ACTIVITY 3

ACTIVITY 5

SYSTEMS AND SUBSYSTEMS

As defined by the *performance specification,* the building becomes a synthesis of several interacting systems and subsystems. In this sense, there are four related, but independently definable subsystem categories that must be successfully assimilated into the total environmental system design. These are: (1) the site systems that provide an environmental context or setting for the building; (2) the subsystems that comprise the basic building enclosure; (3) the subsystems that satisfy environmental and service demands within the interior occupied space; and (4) the subsystems that facilitate the distribution of energy and services to and throughout the building.

FIGURE 3–1.2

THE DIAGRAMMATIC PLAN: SERVICE DISTRIBUTION

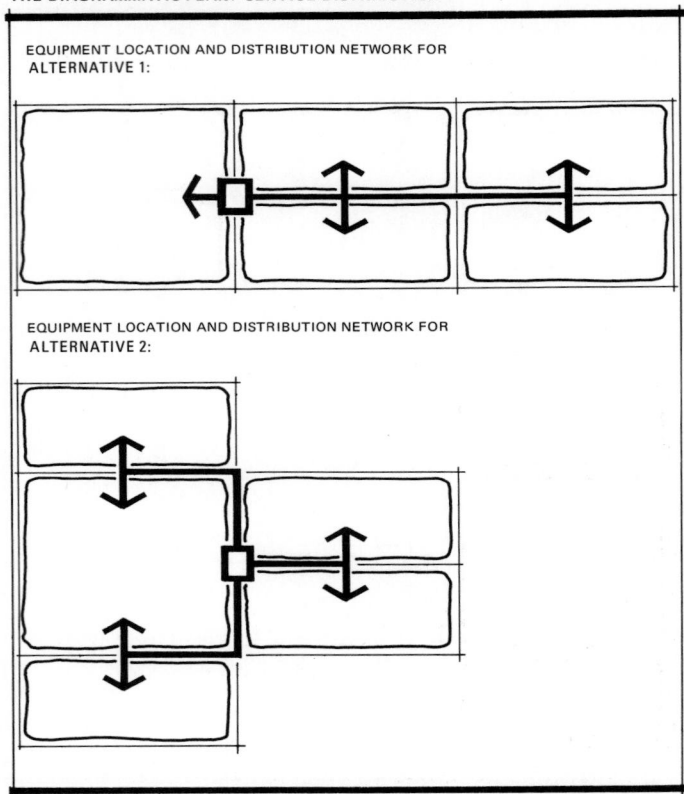

THE SITE

In response to the performance requirements defined in Table 3–1.1, site development can respond to the aspect of environmental comfort in the sense that natural screens and barriers can be utilized for wind control, for sun control, for light control, and for sound control.

Landscape and surface forms

Site vegetation and land forms can influence the immediate thermal environment of the building. These influences generally involve: (1) the diversion of storm winds; (2) the channeling of cooling summer breezes; and (3) sun shading.

Organizational Steps—1

Table 3-1.1 Define the Nature of the Enclosure Shell

Planning Relationship	Organizational Steps	Considerations (refer to discussions in PARTS 1 and 2)
Site-to-area 1	Considering climatic and site conditions, define basic opportunities and limitations regarding placement of interior activities	Re: Daylighting effects
Site-to-area 2		Re: Solar orientation
Site-to-area 3		Re: Prevailing winds and breezes
Site-to-area 4		
Site-to-area 5	Define the essential nature of enclosure form, materials, and construction details	Re: Response to daylighting requirements
		Re: Response to outdoor temperature and humidity conditions
		Re: Response to wind effects and infiltration
		Re: Response to solar radiation effects
		Re: Response to heat storage requirements
		Re: Control of condensation
		Re: Control of internal radiation (MRT)
		Re: Response to noise abatement requirements

Organizational Steps—2

Table 3-1.2 Define the Nature of the Activity Space

Planning Relationship	Organizational Steps	Considerations (refer to discussions in PARTS 1 and 2)
Activity area 1 Activity area 2 Activity area 3 Activity area 4 Activity area 5	Define basic opportunities and limitations regarding room volume and form	Re: Development of daylighting and natural ventilation effects Re: Development of appropriate natural amplification of sound Re: Development of appropriate electronic amplification of sound Re: Development of appropriate reverberation control
	Define basic opportunities and limitations regarding light source selection, reflector-refractor action, brightness and comfort control, housing and protective enclosures, positioning of luminaire elements	Re: Development of appropriate vector influences, spatial order, color *whiteness* Re: Development of appropriate task brightness and contrast Re: Development of appropriate visual comfort
	Define basic opportunities and limitations regarding surface forms, materials, finishes, and assemblies	Re: Response to sonic reflection or absorption requirements Re: Development of appropriate inter-reflection of light Re: Development of appropriate *mean radiant temperature*

Define basic opportunities and limitations regarding air handling elements

Re: Response to environmental temperature and humidity control requirements
Re: Response to air motion and distribution requirements
Re: Air sanitation and odor control

Define basic opportunities and limitations regarding noise-generating devices

Re: Response to limitations in background noise
Re: Development of appropriate noise *screens* to reinforce a sense of privacy

Define the essential nature of the control systems

Re: Lighting switches, dimmers, etc.
Re: Electric service and convenience outlets
Re: Microphone jacks, speaker jacks, etc.
Re: Thermostats, etc.

Organizational Steps—3

Table 3-1.3 Define the Nature of Interior Transitions and Interior Barriers

Planning Relationship	Organizational Steps	Considerations (refer to discussions in PARTS 1 and 2)
Area 1-to-area 2 Area 1-to-area 3 Area 1-to-area 4 Area 1-to-area 5 Area 2-to-area 3 Area 2-to-area 4 Area 2-to-area 5 Area 3-to-area 4 Area 3-to-area 5 Area 4-to-area 5	Locate and define situations where continuity and/or circulation flow is required between adjacent spaces	Re: Development of appropriate continuity or successive contrast in regard to spatial lighting (brightness, color *whiteness*) Re: Development of appropriate continuity or successive contrast in regard to the sonic background Re: Development of appropriate continuity or successive contrast in regard to the thermal and atmospheric background (temperature, humidity, MRT, air motion)
	Locate *barriers* and define their essential nature	Re: Development of appropriate visual privacy Re: Development of appropriate sonic privacy and noise abatement Re: Development of appropriate thermal and/or humidity isolation Re: Development of fire barriers
	Define basic opportunities and limitations regarding *cosmetic* requirements and functions associated with the barrier	Re: Ceiling-floor organization Re: Wall organization

Organizational Steps—4

Table 3-1.4 Define the Nature of the Service Flow

Organizational Steps	Considerations (refer to discussions in PARTS 1 and 2)
Define basic opportunities and limitations regarding the heat source, fuel selection, and related apparatus requirements	Re: Response to heating and humidification requirements
Define basic opportunities and limitations regarding the cooling apparatus requirements	Re: Response to cooling and dehumidification requirements
Define basic opportunities and limitations regarding equipment space requirements and housing characteristics	Re: Location of heating apparatus (including chimney, fuel storage, etc.)
	Re: Location of cooling apparatus (including cooling tower, etc.)
	Re: Location of air handling or fluid control apparatus
	Re: Special characteristics and code limitations
Define basic opportunities and limitations regarding the ventilation system	Re: Location of intakes and exhaust
	Re: Development of appropriate routing
	Re: Development of appropriate air sanitation and composition control (including odor control)
Define basic opportunities and limitations regarding the thermal distribution medium and network	Re: Development of appropriate routing
	Re: Development of appropriate zoning control

FIGURE 3–1.3

LANDSCAPE ELEMENTS FOR THERMAL CONTROL

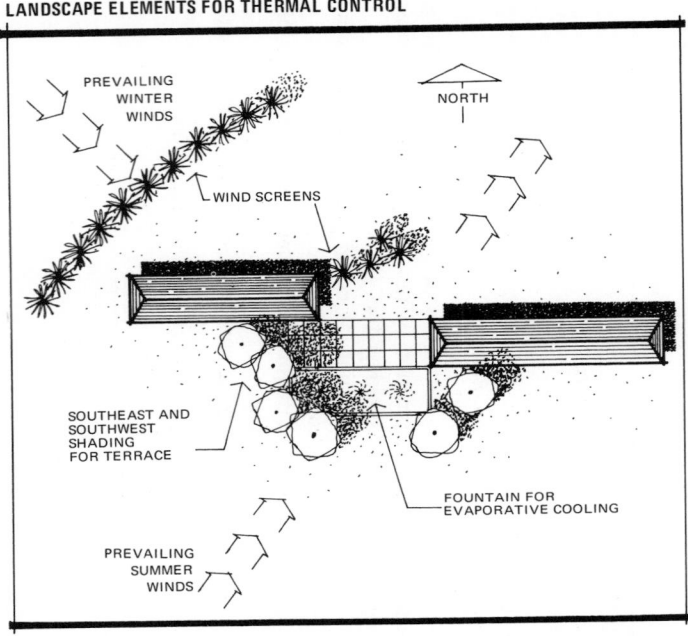

When localized exterior cooling is desired, fountains will provide some evaporative cooling of the air that passes through. For this reason, such water systems should be placed so that moderate-velocity prevailing summer breezes will be induced to pass through the fountain to cool the air prior to its passing into the outdoor living area.

Similarly, wells or a body of water can be useful as a *heat sink*, replacing or supplementing the cooling tower as a means to reject condenser heat in the refrigeration cycle. Fountains can reinforce this action by evaporative cooling of the heat sink.

The typical locations of basic landscape elements for thermal control on an open site are indicated in figure 3–1.3. However, optimum positioning of these elements may vary with local variations in prevailing wind patterns.

In a technical sense, these site factors influence the daytime luminous environment in a somewhat secondary way. Trees and screens that effectively control the thermal influence of solar radiation will also moderate the influences of direct sunlight and sky glare.

When daylight is to be utilized for single-story buildings, this is assisted by providing high-reflectance paved ground surfaces immediately adjacent

Table 3-1.5 Desirable Site Utilization for North-Temperate Climatic Conditions

1. To facilitate maximum exposure to the sun during prolonged winter periods, utilize warm slopes for building sites in colder regions.

2. Where it is desirable to provide natural summer cooling, utilize the lower portion of windward slopes. Furthermore, in order to induce penetration of prevailing summer breezes, openings should be placed to admit ventilation air on the windward side of the building, with exhaust outlets placed on the leeward side.

   As a related factor, in order to facilitate natural internal cooling action during warm periods, minimize blockage of prevailing summer breezes. (Usually this means that dense site screening should be curtailed on the south and southwest exposures.)

3. Wind screening is desirable on the windward side of the building during cold periods. (Usually this consideration applies to the north and northwest exposures.)

4. Utilize evergreens for wind screening purposes. Utilize deciduous trees for sun shading purposes.

5. If possible, locate building so the available fully-developed shade trees will provide shading on the east and west sides of low buildings. Similar considerations apply for the location of outdoor living areas.

6. Paving should be minimized immediately adjacent to the building. Where possible, vegetation should be used in this location to absorb (rather than reflect) solar energy. The critical west and southwest exposures are most likely to produce significant reflected energy during periods of peak solar heat gain.

7. Walks should be shielded from winter winds and summer sun.

to the building. However, this same *reflector* will also reflect significant solar heat into the interior space. For this reason, such reflector surfaces should be used with care, particularly on the west and south exposures.

Noise control

When external noise cannot be muffled at the source, landscape barriers can provide some control within the site. These barriers generally involve either shielding or absorption (or both).

The combination of trees, low foliage, and ground cover provide noise attenuation when significant masses of such absorbing vegetation are involved. Generally, a 500–1000 foot depth of such foilage is required to properly diminish the intensity of normal traffic noises. While relatively thin barriers serve effectively as a visual barrier or sun screen, then, a sonic barrier must be of much more significant dimensions.

FIGURE 3–1.4

DEVELOPMENT OF TOPOGRAPHY FOR NOISE CONTROL

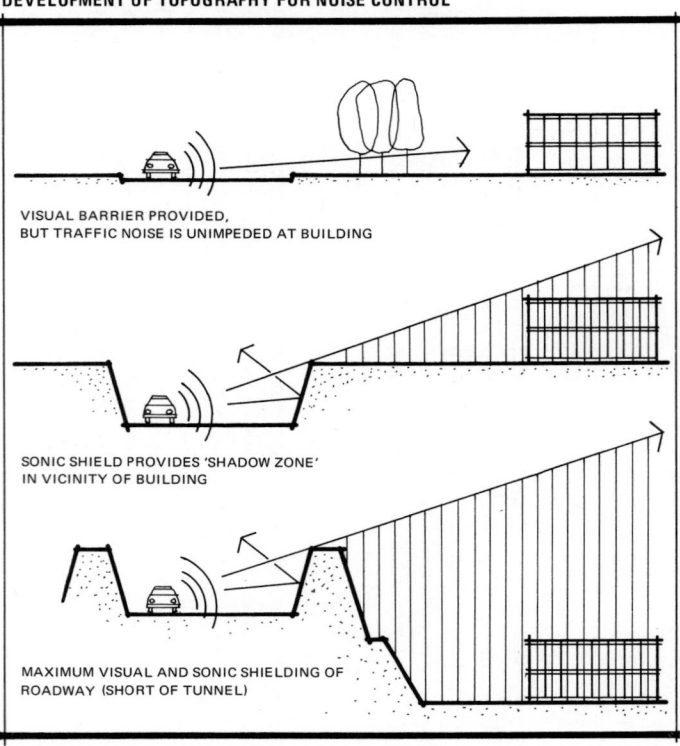

VISUAL BARRIER PROVIDED,
BUT TRAFFIC NOISE IS UNIMPEDED AT BUILDING

SONIC SHIELD PROVIDES 'SHADOW ZONE'
IN VICINITY OF BUILDING

MAXIMUM VISUAL AND SONIC SHIELDING OF
ROADWAY (SHORT OF TUNNEL)

More effective control may be provided with shielding (see Figure 3–1.4). In order for such a shield to be effective, it must have significant mass and it must be impervious to air flow. The shield must also be of sufficient height to screen the line of sight between the noise source and the receiver. This technique casts a *sonic shadow* over areas bordering the noise source. This deflection of sound creates an area of reduced noise intensity (up to 25 db) within the *shadow zone*.

Site service networks

The building environmental and service systems depend on the provision of adequate site systems for electric power, water supply, gas, communica-

FIGURE 3–1.5

PRESERVATION OF SERVICE ACCESS

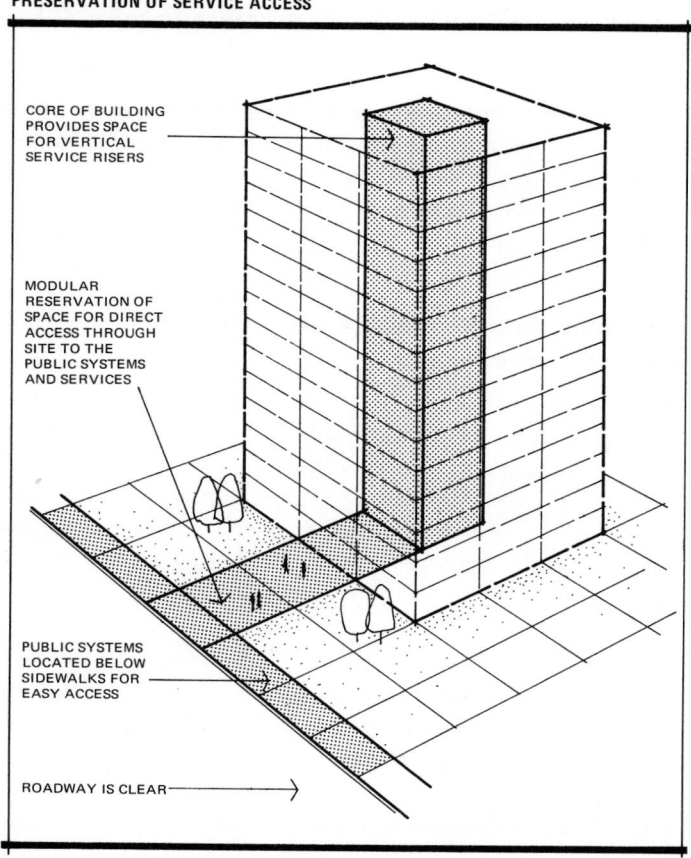

CORE OF BUILDING
PROVIDES SPACE
FOR VERTICAL
SERVICE RISERS

MODULAR
RESERVATION OF
SPACE FOR DIRECT
ACCESS THROUGH
SITE TO THE
PUBLIC SYSTEMS
AND SERVICES

PUBLIC SYSTEMS
LOCATED BELOW
SIDEWALKS FOR
EASY ACCESS

ROADWAY IS CLEAR

tions, and liquid disposal. These utilities and services can be distributed in walk-through tunnels that are fully accessible for maintenance. However, this method of distribution tends to be costly and is generally limited to compact projects.

More generally useful are concrete or tile trenches and metal conduit enclosures, both placed under the sidewalk (see Figure 3–1.5). The paving can then be designed to be lifted so the piping is accessible as required for maintenance and future connections. This access should be available without disturbing the landscaping.

THE ENCLOSURE SYSTEM

Immediately related to the aspect of site development is the development of the enclosure system that defines the external character and form of the building.

In one sense, the enclosure serves as a *barrier* to protect the interior space from hostile or inappropriate external influences. At the same time, the external environment contains many favorable influences; so the enclosure may serve more precisely as a selective *filter* which allows these favorable influences to penetrate. In this sense, the appropriateness of the enclosure system depends on its success in intercepting, modifying or otherwise regulating the penetration of external influences into each activity-oriented interior space.

Decisions regarding dominance of natural or mechanical influences

As interior activities are studied, it is possible to make decisions regarding the appropriateness of various influences in the natural (exterior) environment. At this point, these influences can be either utilized or excluded through the methods discussed in Chapters 2–1 and 2–3.

Tables 3–1.7 and 3–1.8 deal with the general situations and conditions in which natural and electrical-mechanical alternatives become relevant.

Thermal performance of building forms

When developing the enclosure form itself, it is significant that each face of the building is subject to somewhat unique climatic influences related to wind, sun, and conductive interchanges with the air mass. These influences are discussed in Chapter 2–3, and the more significant solar patterns are briefly summarized in Figure 3–1.6.

Table 3-1.6 The Enclosure System

Subsystems	Performance Functions
The roofing subsystem	1. Control solar heat penetration through openings
The exterior wall subsystem	2. Control transmission of light through openings
The foundation-floor subsystem	3. Control heat conduction through the assembly
	4. Control condensation within the assembly
	5. Control the *mean radiant temperature* effect of the interior faces of the enclosure
	6. Control the transmission of external noise
	7. Control the penetration and exhaust of ventilation air

Table 3-1.7

Natural Light Implies:	Electric Light Implies:
1. Narrow rooms Rooms with high ceilings Ability to utilize roof openings and related shielding elements	1. Deep, wide rooms Rooms with low ceilings Interior spaces with no openings to the outside
2. Activities where observation of external areas is desirable as an environmental or functional influence (NOTE: This factor is independent of the lighting function, but has obvious implications in terms of brightness distribution and color of light in the room.)	2. Provision of illumination supplementary to daylight (NOTE: Which further implies a need to develop electric patterns that are generally compatible with daylight, both in the sense of brightness distribution and color quality.)
	3. Provision of illumination in rooms where glass is impractical or undesirable: • because of location in a building • because of thermal exposure problems • because of a need to minimize external visual or sonic distraction • because of special problems associated with control of light distribution or color of light
Generally to be used: • where diffuse lighting is acceptable during most of the day • where nonuniform distribution and noncontrollable variations in light intensity, color, and directional quality are desired or acceptable	Generally to be used: • where control over vector influences, spatial contrast, and color of light is required • where some degree of consistency and predictability is required, and periodic variations in natural light cannot be tolerated (as in spaces intended for sustained visual work)

• where landscaping, adjacent buildings, or integrated solar screens will adequately control glare

• where dimming and selective switching are required
• where special effects or special *mood* requirements are involved (as in a theater)

Table 3-1.8

Natural Ventilation Implies:	Mechanical Ventilation Implies:
1. Narrow buildings with adequate openings which permit a natural flow of air through the building (i.e., breezes are permitted to enter through the windward side, penetrate through the building, and exit through the leeward side)	1. Deep or complex building forms which prevent or inhibit a natural flow of air through the building (Including buildings without adequate openings)
Generally to be used:	Generally to be used:
• where external air conditions are acceptable (re: cleanliness, temperature, humidity, etc.) • where openings directly into the occupied space are acceptable • where nonuniform distribution of air and non-controllable quality characteristics are desired or acceptable	• where special conditioning of air is required (re: filtering, temperature control, humidity control, etc.) • where openings directly into the occupied space cannot be tolerated for visual, sonic, or thermal reasons • where some degree of consistency and predictability is required, and periodic variations in air flow and air quality cannot be tolerated

FIGURE 3–1.6

SOLAR LOADS THROUGH OPENINGS

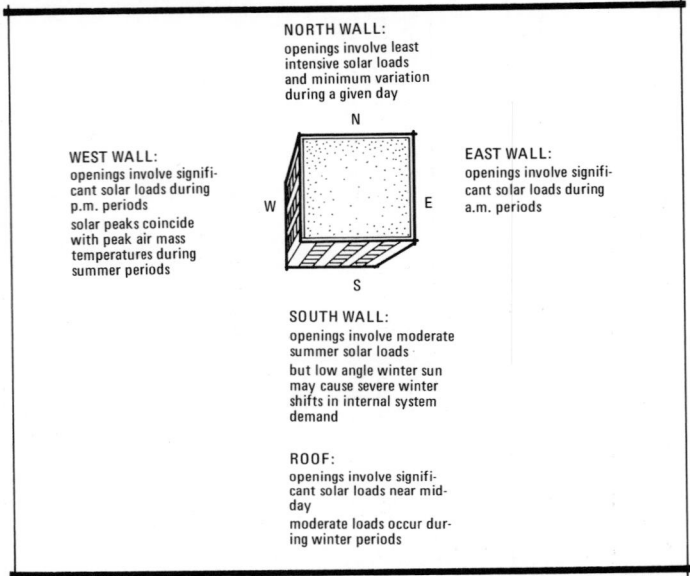

NORTH WALL:
openings involve least
intensive solar loads
and minimum variation
during a given day

N

WEST WALL:
openings involve signifi-
cant solar loads during
p.m. periods
solar peaks coincide
with peak air mass
temperatures during
summer periods

W   E

EAST WALL:
openings involve signifi-
cant solar loads during
a.m. periods

S

SOUTH WALL:
openings involve moderate
summer solar loads
but low angle winter sun
may cause severe winter
shifts in internal system
demand

ROOF:
openings involve signifi-
cant solar loads near mid-
day
moderate loads occur dur-
ing winter periods

Performance of compact and finger-type building plans But beyond the implications of differing exposures associated with individual
building faces, a more general relationship exists between the essential
form of the building and the operating performance of balancing mechanical devices.

For example, a *compact* building form is one that encloses a maximum
floor area with minimum perimeter footage. (As an example, a square is
the most compact rectangular form.) This minimizing of the interior-
exterior contact surface will generally improve the thermodynamic characteristics of the enclosure in both cold and hot-arid situations. However,
compact building forms tend to be somewhat limited in their ability to
respond to moderate and humid climatic conditions, when ventilation flow
is decisive in maintaining occupant comfort. As a result of this limitation,
these building forms will present more extensive cooling demands during
moderate and humid periods.

Irregular *finger-type* or decentralized building plans require a greater
perimeter footage to enclose the same floor area. These forms will tend
to permit a better natural interaction during moderate and humid periods
because ventilation air can generally be led to circulate naturally through
the narrower building sections (by cross-flow). However, these finger-type
forms offer a poorer response during cold periods because of the greater

interior-exterior conduction surface. As a result, these building forms will present more extensive heating demands during winter periods.

Since north temperate regions tend to evolve a variety of external conditions during the various seasons of the year, then, both of these basic building form alternatives will be somewhat limited in ability to respond naturally to thermal influences during some portions of the year. For this reason, a balancing mechanical system must become an integral part of the concept; and the relative size and significance of individual mechanical components will vary with the building form.

Comparative performance of building shapes   Figure 3–1.7 summarizes heat gain and heat loss performances for several typical building configurations.

Notice that as the building becomes larger (vertical or horizontal expansion), the heating and cooling systems will operate more efficiently because the relative influence of heat transfers through roof or wall surfaces is reduced.

Also, when the building shell involves minimum window openings, conductive heat transfer is a dominant thermal influence. Therefore both heating and cooling demands are minimized with compact building shapes. On the other hand, when significant glass areas are involved, structures that are elongated in an east-west direction may tend to minimize cooling demand. This is particularly true when internal heat sources are negligible.

However, this latter conclusion tends to be complicated somewhat when a major internal heat load is present (such as a high-level electric lighting load). Since orientation along the east-west axis still leaves the south face exposed to significant direct radiation on clear winter days when the solar altitude is low, winter solar heat (in addition to significant lighting heat) can produce overheated interior conditions, even on a very cold day. For this reason, glass in the south wall of commercial or institutional buildings should be used with restraint, unless adequate shading is provided.

Related internal planning relationships

The previous comments have referred primarily to the nature of exterior walls as filters for regulating the penetration of external environmental influences. However, as a related technique, a selective placement of activities and rooms can become useful as a reinforcing device in both sonic and thermal development.

Insulation from noise influences   The enclosure shell should insulate the interior space from external noises associated with traffic, aircraft,

FIGURE 3–1.7

THERMAL PERFORMANCE OF TYPICAL ENCLOSURE CONFIGURATIONS

		PEAK WINTER HEAT LOSS	PEAK SUMMER HEAT GAIN	SUMMER AIR HANDLING
		BTU / SQUARE FOOT OF FLOOR AREA		CFM/ SQUARE FOOT
VARYING SHAPE AND ORIENTATION (CONSTANT FLOOR AREA)	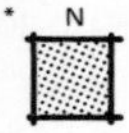 N (square)	−32	+56	1.9
	 N (vertical rectangle)	−36	+64	2.2
	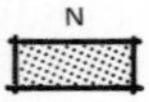 N (horizontal rectangle)	−36	+55	1.8
VARYING VERTICAL VOLUME	1 FLOOR	−42	+64	2.2
	5 FLOOR	−32	+56	1.9
	20 FLOOR	−30	+55	1.8
VARYING HORIZONTAL VOLUME	50 × 50	−46	+64	2.2
	100 × 100	−32	+56	1.9
	200 × 200	−25	+48	1.5
VARYING GLASS OPENINGS	66% GLASS	−41	+67	2.4
	33% GLASS	−32	+56	1.9
	NO GLASS	−23	+46	1.4

rail activity, stationary machinery, playground activity, etc. In this sense, the shell should be sealed to prevent air-borne noise penetration, and it should provide suitable mass or structural discontinuity to inhibit the penetration of external noises (see Chapter 2–2).

When the enclosure itself may be inadequate to provide a completely effective barrier, critical interior areas should be further shielded from troublesome external noise sources. As an example, the bedrooms of a home can be shielded from external road noises and playground noises by using other less critical interior spaces as separating *buffers* (see Figure 3–1.8).

FIGURE 3–1.8

SONIC SHIELDING OF QUIET INTERIOR AREAS

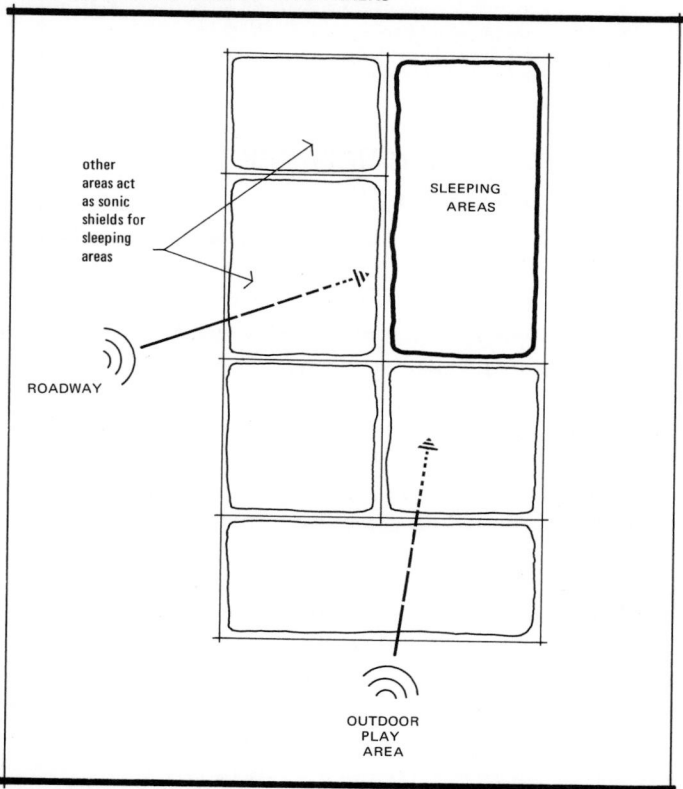

Insulation or exposure of thermal influences Additional layout considerations evolve from thermal influences. General recommendations are summarized in Table 3–1.9 and among the final items in Table 3–1.10.

This latter table also provides a more general summary of thermal considerations associated with the enclosure.

Table 3-1.9 Recommended Solar Orientation for Various Rooms

	Orientation							
	N	NE	E	SE	S	SW	W	NW
Work space with exterior exposure	X	X						X
Recreational and lounge areas with exterior exposure				X	X	X	(X)	
Courtyards and terraces			(X)	X	X	X	(X)	

NOTE: The American Public Health Association committee on the hygiene of housing has recommended: "at the winter solstice, at least one-half of the habitable rooms of a dwelling should have a penetration of direct sunlight of one-half hour's duration during the noon hours when the sun is at its maximum intensity."
(This recommendation refers primarily to the germicidal action of solar radiation.)

THE SERVICE DISTRIBUTION NETWORK

The initial function of the mechanical service system is to compensate for environmental deficiencies or excesses caused by inconsistent interaction between the building shell and the surrounding natural environment. In the temperate regions of North America, the required service distribution networks and related process equipment are often relatively extensive, and they tend to make rather extensive demands on the physical form of the building.

As a result, these service networks are somewhat expressive of the nature of the building. They define the flow of energy and services; and in this sense, the network should be a disciplined and logical expression that facilitates both accessibility and flexibility for change.

Essentially, then, the concept of the building should reflect a number of system decisions that can be defined within the context of the diagrammatic plan (see Figure 3–1.2 and Table 3–1.4). How are the services distributed vertically? How are they distributed horizontally? Is there to be a *centralized* or *decentralized* distribution network? And finally, what space and linkages are required for the major process equipment?

Mechanical service space

Most buildings are composed of three types of interior space. One type is the useful or functional space that is available for the essential human

Table 3-1.10 Desirable Characteristics for Exposed Enclosures in North Temperate Climates

1. In terms of the utilization of mechanical heating and cooling capacity, *compact* building forms (i.e., those with a minimum *shell area to enclosed area* ratio) will tend to be more optimum energy-conserving building shapes when the enclosure is exposed to extreme seasonal conditions (cold or hot-arid). However, significant mechanical ventilation and cooling of interior core area will be required during moderate and humid periods because *compact* building forms tend to be limited in their ability to interact naturally with the external environment.

2. In contrast with the previous comments relating to *compact* building forms, penetration of natural light, natural cross-ventilation, and free air flow through interior spaces all imply the development of narrow building elements and *open* plan interior partitioning.

3. When only minimum to moderate internal heat source concentrations are present in the interior, building shapes that are elongated along the east-west axis are generally preferable in that they tend to minimize solar heat gain during peak summer periods.

4. Utilize light to medium exterior finishes for exterior walls and roof areas. Dark absorbent colors should be restricted to accents or to areas that are shielded from the summer sun.

5. When possible, shading devices should be external—and they should be detached from the structure in the sense that they should be exposed to maximum cooling by wind convection.

6. Wall and roof construction should have good insulation characteristics, be nonporous, and be resistant to freeze-thaw action. Construction should resist moisture penetration. (These characteristics tend to be most critical on the windward north and west exposures.)

7. Where appropriate, roof construction in snow areas should utilize simple sloped roof forms to facilitate snow removal by wind action. Simple roof forms will minimize snow accumulation and ice-filled gutters.

8. When buildings are placed on open sites, plan development should generally be open to the south or southeast, and should be relatively closed to the north and west (to screen storm winds and late afternoon sun). Therefore, window areas on the north and west exposures should be somewhat limited.

9. When buildings are placed on open sites, noninhabited spaces should generally be placed in the west and southwest portions of the plan to serve as insulators against afternoon solar effects.

10. Courtyards and outdoor lounge areas should be oriented toward the south in order to maximize the use of these areas as an extension of interior spaces during periods of moderate to cool weather. Shading of the west sun is desirable during summer periods.

11. Interior heat-producing and humidity-producing areas should be separated from other occupied areas.

activities that take place within the building. This type will include office space, classroom space, lounge space, conference rooms, eating space, etc. —and in the sense of the current discussion, these spaces provide the predominant framework for placement of terminal air diffusers, luminaires, acoustical absorption, etc. (to be discussed in the succeeding section).

The second type of space is the major human circulation and exit routing that provides physical access to and from human activity areas. This will include entrance lobbies, stairs, elevators, corridors, etc., and will probably also include personal or building services that are immediately accessible from the circulation route (areas such as public toilet rooms and janatorial spaces). In the sense of the discussion in this section, this human circulation path provides a potential framework for the organization and placement of centralized distribution networks to and from the functional spaces. Duct and other service routes can parallel horizontal corridors and vertical stair towers.

A third type of space is *mechanical service* space. This will include some areas that require intermittent human access (such as equipment rooms, fan rooms, etc.) and other areas that are somewhat remote from immediate human access (such as vertical and horizontal trunk duct space). As a category, however, the placement and sizing of mechanical service spaces are generally dependent on equipment needs rather than human needs.

Each of these three space categories is essentially separate and definable within the building design.

Layout and expansion of mechanical service spaces

Table 3–1.11 summarizes some basic space allowances that are useful for estimating purposes during the initial schematic design phase.

When significant future expansion is anticipated, equipment and plant design should be based on carefully defined increments of growth. These increments (equipment modules) can then be installed in stages as required, utilizing previously reserved areas and a common distribution network.

The initial stage of a central mechanical service plant, for example, may include two equipment modules; one to provide basic capacity, and one to provide 100% standby capacity. As the building size is doubled, a third equipment module is added. In this way, a 100% growth in load requires only a 50% expansion of service capacity (because one equipment module is always available as standby to take over for one of the other two). As this process continues in future expansions, an efficient system of service units evolves that is capable of responding at various capacity levels to meet varying load conditions.

Table 3-1.11 Mechanical Space Allowances for Use in Preliminary Planning

Centralized mechanical equipment rooms	Allow approximately 5–8% of the gross floor area for refrigeration equipment, heating devices, and related pumps.
	When positioning boiler and chiller devices, allow adequate space immediately adjacent to the equipment to permit removal and maintenance of the tubes. Generally this space allowance approximates the length of the device.
	Allow equipment room ceiling heights of approximately 13–18 ft.
	Allow approximately 2% of the gross floor area for air handling components.
	Consider thermal and acoustical insulation requirements when developing wall, ceiling, and floor details for all equipment rooms.
Cooling towers	Allow approximately 1 sq ft of roof area for each 400 sq ft of gross building floor area.
	Height allowances range from approximately 13 ft (for small buildings) to approximately 40 ft (for very large buildings or building groups).
	Structural load allowances range from 120–200 psf with water.
	Allow approximately 4 ft of access space below the unit for access and piping.
	Consider acoustical isolation requirements when developing adjacent roof and wall details; also consider the need for resilient cushioning to prevent structural transmission of vibrations.
Chimneys	For gravity systems, place the natural draft chimney 100 ft or more from fan room air intakes and cooling tower intakes.
	Forced-draft chimneys can generally be placed in closer proximity, provided exhaust air flow is directed away from air intakes.
Split package mechanical equipment	Typical 5–20 ton limit per unit.
	Locate air-cooled condenser units near the perimeter because of difficulties in overcoming duct friction.
	Remote interior evaporator is linked to the condenser with piping:
	• Maximum horizontal distance to interior evaporator: Approx. 60 ft.
	• Maximum vertical distance to interior evaporator: Approx. 30 ft.
	(Therefore, when multiple units are to be used, consider (1) 5-floor vertical zones, with the mechanical unit located at the center floor; or (2) utilization of one unit to serve each moderate-sized floor).

Table 3-1.11 (Continued)

Roof-top, multi-zone packages	Typical 5–60 ton limit per unit. These are self-contained heating, cooling, and air handling devices (excluding ducts). Height allowances range from approximately 5 ft (for smaller units) to 10 ft (for larger units). Roof area allowances range from approximately 4 sq ft per ton of capacity (for larger, more efficient packages) to 6 sq ft per ton (for smaller packages).
Centralized mechanical service zones	One centralized mechanical equipment room will service approximately 8–20 floors of a high rise building. The more floors served, the larger the duct shafts and the greater the mechanical equipment room volume at each single location.

Air duct allowances	Horizontal duct allowances: (for estimating *drop ceiling* cross sections for corridors, etc.)
	  Allow approximately 1–2 sq ft of cross section for supply ducts for each 1000 sq ft of occupied space.
	  (This will vary, depending on the velocity of air in the trunk, i.e., higher velocity air flow requires smaller ducts. The allowance will also vary with the complexity of the duct layout and with the air change rate required, i.e., simplified duct layouts and reduced heat gains will reduce duct size requirements.)
	  The supply duct allowance must be duplicated for the return duct system.
	  Consider the fact that return air velocities may be lower than the supply requiring larger duct allowances.
	Vertical shaft allowances:
	  Shaft allowances for a given floor and for a group of floors will be based on the same supply and return allowances considered above.
	  Shaft and duct allowances may need to be increased 2 times or more to compensate for air flow inefficiencies such as excessive elbows, angles, boot fittings, duct take-offs, etc. For this reason, complex and irregular duct patterns should be avoided.

The mechanical service zone

Mechanical service zones are those limited portions of a building that can be defined as a *detachable* or semi-independent mechanical operating unit. As such, it becomes one of the organizing elements in building planning.

Generally, these zones are repeatable units, such as a floor of a building, a group of floors, or some other definable structural segment. In each case, the zone is served with electric power, communications, plumbing, and the H.V.A/C services necessary to facilitate the intended activities within that zone.

For some of these services, the zone may be self-sustaining in the sense that package processing equipment is located integrally within the zone. In other cases, all services are brought to the zone from some centralized remote equipment location. But in either case, each mechanical service zone involves: (1) a service trunk and entrance that serves as a point of access and control, and (2) a comprehensive internal distribution network that provides for internal service access to all parts of the zone.

The definition of this service zone (or zones) is an important aspect of planning because it establishes a discipline within which the various electrical and mechanical advisors can work in a sense of harmony with the architectural design. Without such a definition, this aspect of building design tends to remain fragmented and undisciplined.

Centralized distribution networks

The service distribution networks must physically relate to each of the three types of interior space (functional space, human circulation or access space, and mechanical equipment areas). This fact can become a useful discipline in network planning; because when circulation and access evolve around the concept of a core-corridor system, one logical distribution path for vertical and horizontal mechanical services will parallel the vertical and horizontal human circulation route.

According to this theory, the access core that provides a continuous vertical right-of-way for elevators and stairs will also provide a manageable vertical route for major liquid, air, and electric power distribution risers. Mechanical equipment and devices will be clustered in a convenient manner near this core (see Figure 3–1.9).

Similarly, the major horizontal mains or trunks will parallel the major corridor systems (as in a drop ceiling). Feeders can then run off from these trunks as necessary to serve individual or clustered rooms in a manner not unlike the way that doorways provide access to these same

FIGURE 3–1.9

THE CENTRALIZED MECHANICAL DISTRIBUTION NETWORK

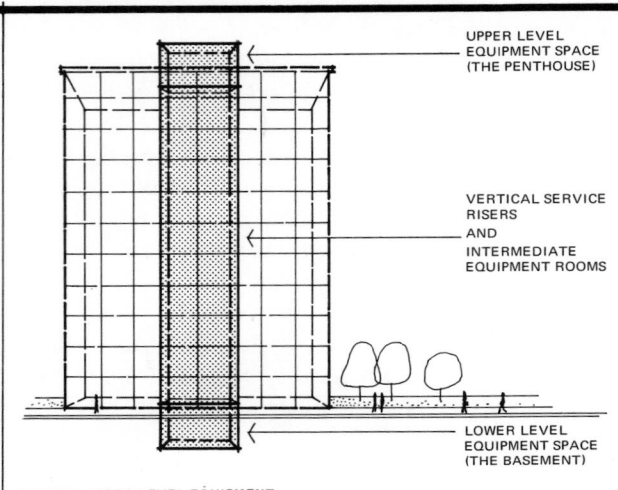

UPPER LEVEL
EQUIPMENT SPACE
(THE PENTHOUSE)

VERTICAL SERVICE
RISERS
AND
INTERMEDIATE
EQUIPMENT ROOMS

LOWER LEVEL
EQUIPMENT SPACE
(THE BASEMENT)

TYPICAL UPPER LEVEL EQUIPMENT

Cooling tower
Alternative chiller equipment location
Alternative heating equipment location
Alternative central fan room location
Elevator penthouse, water tanks, etc.

TYPICAL CORE ELEMENTS

Vertical air supply and air return riser mains
Fluid risers which link upper level and lower level equipment components
Electric and plumbing service risers

Vertical service routing parallels the human circulation route---i.e.
the elevator shaft, the stair tower, etc.

The core includes toilet rooms, janitor service spaces, and other
'wet' service spaces

INTERMEDIATE EQUIPMENT (IF ANY)

De-centralized chiller, heating, and air handling locations as necessary
at intermediate levels

TYPICAL LOWER LEVEL EQUIPMENT

Alternative chiller equipment location
Alternative heating equipment location, and related fuel storage
Alternative central fan room location
Electrical and plumbing service rooms

HORIZONTAL SERVICE NETWORKS (TYPICAL FLOOR)

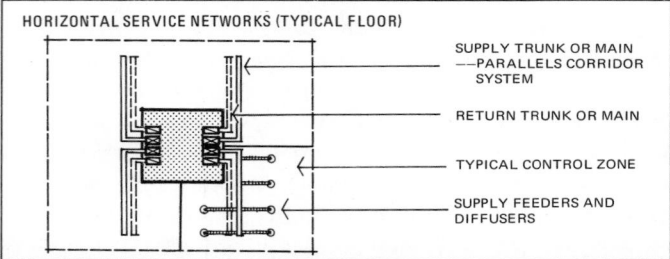

SUPPLY TRUNK OR MAIN
---PARALLELS CORRIDOR
SYSTEM

RETURN TRUNK OR MAIN

TYPICAL CONTROL ZONE

SUPPLY FEEDERS AND
DIFFUSERS

FIGURE 3–1.10

TYPICAL PLACEMENT OF SERVICE CORES

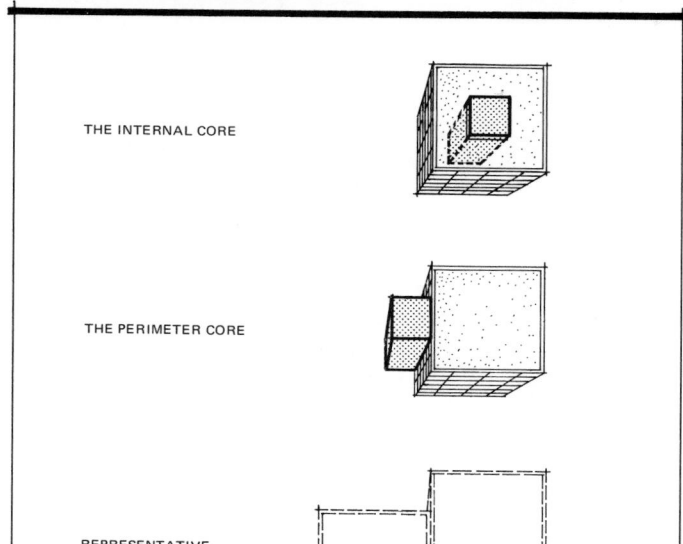

THE INTERNAL CORE

THE PERIMETER CORE

REPRESENTATIVE
ARRANGEMENT TO
FACILITATE FUTURE
EXPANSION

rooms from the corridors. This procedure will insure an organized and accessible right-of-way to both near and remote activity areas.

Decentralized distribution networks   An alternative routing of services can parallel the structural system. Because of the inherent decentralization of the structural columns, this concept will naturally produce a decentralized vertical distribution network.

According to this theory, vertical routing is contained within the column, while horizontal routing of mains and trunks is contained within the major structural beams. In such cases, the need for service space will probably lead to a somewhat more complex definition of the structure than is likely when only simple structural loading is involved. For example, accessible spaces may be defined by double beams and columns, or structural members may serve a dual capacity as a riser shaft or duct (see Figure 3.1.12).

FIGURE 3–1.11

TYPICAL MASSING OF MECHANICAL SERVICE SPACES

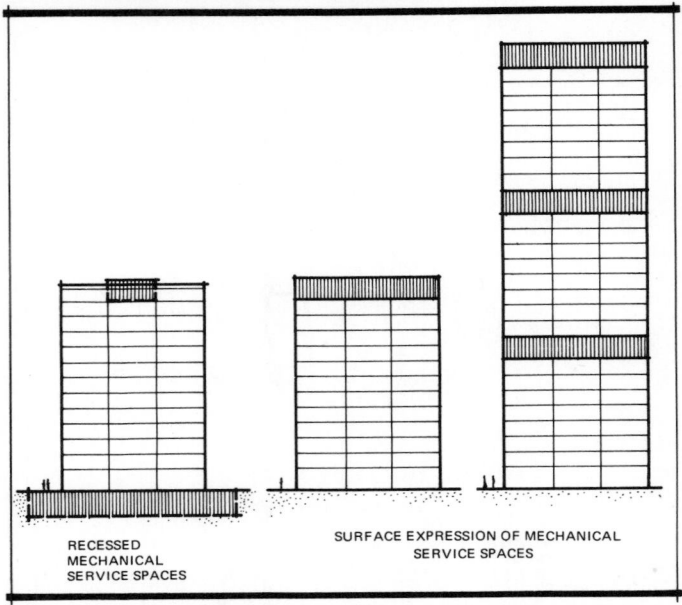

RECESSED
MECHANICAL
SERVICE SPACES

SURFACE EXPRESSION OF MECHANICAL
SERVICE SPACES

THE COMPREHENSIVE INTERIOR ENVIRONMENTAL SYSTEM

The basic activity unit is generally a room or a similar spatial subdivision. The specific needs of this room will vary according to the nature of the occupancy or activity. But in order for the space unit to function successfully, it must perform a number of environmental and service functions that are summarized in Table 3–1.12.

If the interior spatial unit is to function in a successful way, all aspects of the sensory environment (luminous, thermal, and sonic) must be satisfied *concurrently* as integral parts of a total environmental setting or background. While the individual disciplines of light, heat, and sound can be discussed as somewhat fragmented concepts, then, the actual interior spatial unit must represent a synthesis.

System components and assembly

Comprehensive interior systems may be developed and detailed individually for a specific building project. They may also reflect contemporary methods of industrial production and assembly, particularly for high production *vernacular* building types such as office buildings, schools, stores,

FIGURE 3–1.12

THE DE–CENTRALIZED MECHANICAL DISTRIBUTION NETWORK

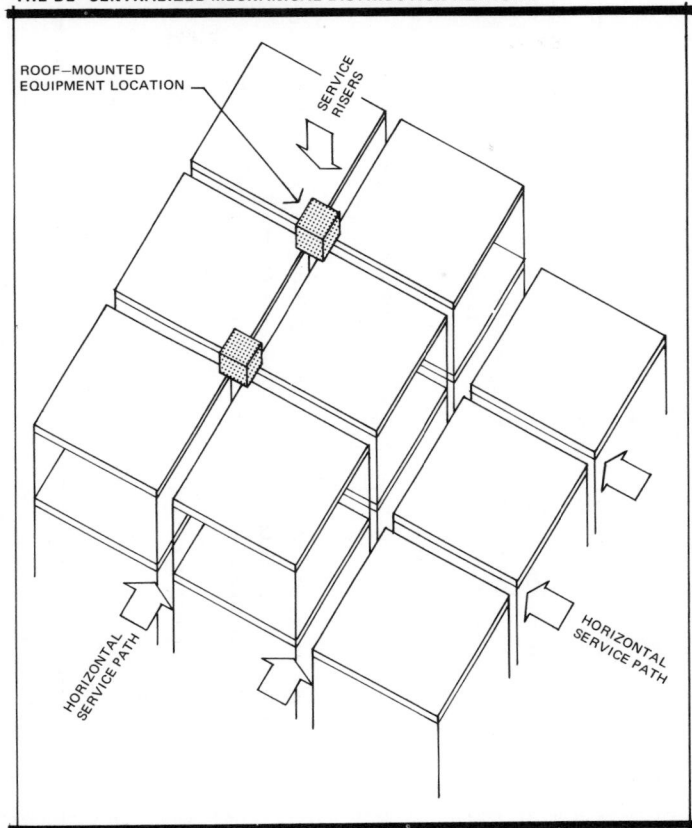

laboratories, etc. Each of these is generally characterized: (1) by the repetitive nature of the interior spaces, and (2) by the requirement of flexibility to facilitate change and growth.

In response to the needs of such building types, then, there is a contemporary search for effective direction and integrity in four related areas. These are:

(1) the use of assembly line techniques to provide for subassembly of components at a remote location, and then rapid integration of effective systems on the job site.

(2) the use of various modular approaches in an attempt to solve the distribution problems associated with industrialized components; the

Table 3-1.12 The Interior System

Subsystems	Performance Functions
The partition subsystem The ceiling-floor subsystem	1. Control air-borne sound transmission between spaces. 2. Prevent structural-borne sound transmission between spaces. 3. Provide appropriate sound control within the occupied space (i.e., absorption or reflection). 4. Provide appropriate intensity, color, and distribution of light in the occupied space (i.e., light sources, reflectors, lenses, brightness control devices). 5. Provide appropriate light reflection characteristics within the occupied space. 6. Provide air or water distribution for thermal processing within the space (i.e., ducts, piping, valves, dampers, mixing boxes, etc.). 7. Provide for electrical distribution and control (i.e., wireways, outlets, switches, etc.). 8. Provide for plumbing service distribution and control (if applicable for *wet* spaces). 9. Provide for distribution of communication services (i.e., wireways, outlets, etc.).

current need to manufacture and stock building components without knowledge of who will buy it or where it will be used.

(3) the solution to the problems associated with joining separate components and subassemblies, and the reduction of special skills or tools required for field assembly.

(4) the fundamental need to provide flexibility for spatial variation.

Industrialization   The techniques of mass production offer the possibility (if not the immediate realization) of precision components and low building costs, if the building process can be organized to utilize the unique potential of the production line while recognizing its limitations.

This requires that a successful building approach (for interior environmental systems) be developed to utilize repetitive units produced on a standard assembly line; or it requires that production line techniques be utilized to permit economical unit variations in form, dimension, texture, and color. In either case, industrialization requires: (1) a minimizing of the number of different components or assemblies required, and (2) a minimizing of the physical steps required of both factory and field labor.

Modular development   The nature of assembly-line procedures has produced considerable experimentation in the techniques of dimensional coordination and modular repetition.

The *standard dimensional module* is an attempt to provide a horizontal and vertical dimensional framework (planning grid) through which the various building design disciplines can correlate components and assemblies. This approach utilizes structural members of a consistent and common length, partition units and electrical-mechanical elements of related size, etc. Figures 3–1.13 through 3–1.19 relate to this basic approach to industrialized building design.

The *human-use module* is a spatial approach rather than one oriented toward component assembly. This module could be a completely prefabricated bathroom, kitchen, office, or classroom; possibly shipped complete to a job site for incorporation as a finished product. It includes all elements necessary to perform a given task or activity. (In theory, an automobile or an aircraft are examples of mass-produced, regularly-variable human-use modules.)

The Mechanically Coordinated Module.   The term *mechanically coordinated module* refers to a standard spatial unit that includes lighting, air supply, air return, and appropriate sound control, together with the necessary sanitary, communication, and power services. (See Figure 3–1.14.)

Each individual environmental and service component is dimensionally

Table 3-1.13 Potential Advantages of Industrialized Interior Systems

1. Utilization of pretested and coordinated components facilitates the development of predetermined quality levels and performance standards.

2. Utilization of dimensional and performance disciplines facilitates the ability to preselect compatible structural components, partition-enclosure components, and environmental control components—and this facilitates a reduction in field design and engineering time.

3. Predictability of erection procedures facilitates more accurately determined cost projections and bidding procedures.

4. Utilization of standardized components facilitates speed or delivery and improved construction scheduling.

5. Utilization of pretested and coordinated assemblies facilitates erection procedures, provides improved component consistency, and generally provides improved installation economies over conventional construction of comparable quality.

6. Emphasis on production compatibility and interchangeability facilitates flexibility for spatial rearrangement and modification as activity requirements evolve and change.

FIGURE 3–1.13

THE SYSTEM MATRIX GRID

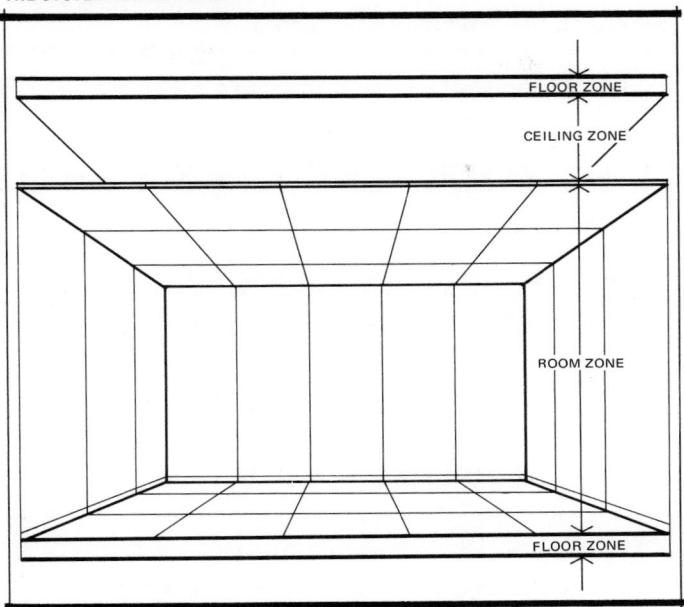

related to the modular floor, wall, and ceiling systems; and the combination must be compatible with the structural and space furnishing systems. The resulting assembly may also perform a cosmetic function by correlating (and possibly concealing) the somewhat complex assembly of components and networks.

The mechanically-coordinated module, then, is the basic self-contained unit of repetitive interior assembly. It should be sufficiently flexible to provide maximum design freedom for the initial designer or the future tenant. And yet, the systems should be sufficiently disciplined in a technical sense to insure that the resulting rooms can be made to function in a logical and efficient way.

When this approach is effectively purseud, the mechanically-coordinated module becomes a basic subsystem in interior planning. Any combination of modules can be isolated and identified as an independent room; and the resulting space will automatically include the capability to provide for luminous, thermal, sonic, and other service needs (see Figures 3–1.18A thru 3–1.18C).

Assembly of components   A third factor in the analysis of interior environmental systems is the problem of on-the-job assembly. This involves

FIGURE 3–1.14

THE PLANNING MATRIX:
SPACE ALLOCATION IN THE CEILING ZONE (STEEL)

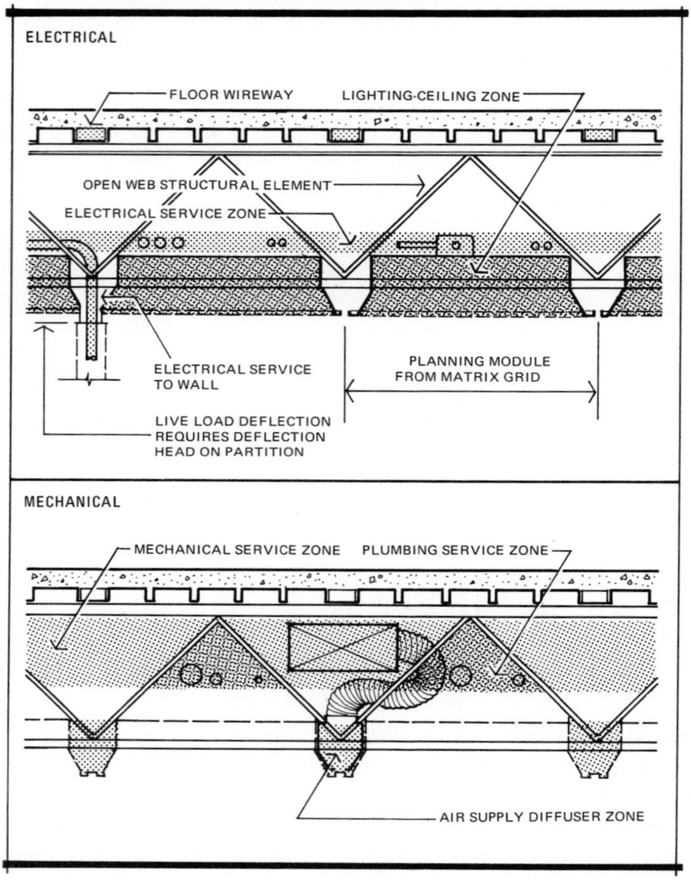

ELECTRICAL

FLOOR WIREWAY   LIGHTING-CEILING ZONE

OPEN WEB STRUCTURAL ELEMENT

ELECTRICAL SERVICE ZONE

ELECTRICAL SERVICE
TO WALL

PLANNING MODULE
FROM MATRIX GRID

LIVE LOAD DEFLECTION
REQUIRES DEFLECTION
HEAD ON PARTITION

MECHANICAL

MECHANICAL SERVICE ZONE   PLUMBING SERVICE ZONE

AIR SUPPLY DIFFUSER ZONE

the relationship between adjoining elements. Methods of joining become
critical considerations, and potential areas of controversy and conflict.

In this sense, a desirable objective for any industrialized building section
is an extremely simple and well-defined method of joining; one that reduces to a minimum the special skill or tools required for field assembly.
Complex connections, excessive maneuvering, or field alterations quickly
submerge the economic advantages of industrialization.

Component and Joint Tolerances.   A basic problem in industrialization
(and perhaps the one most difficult to solve) is that the size and shape

FIGURE 3–1.15

THE PLANNING MATRIX:
TYPICAL AIR HANDLING DEVELOPMENT

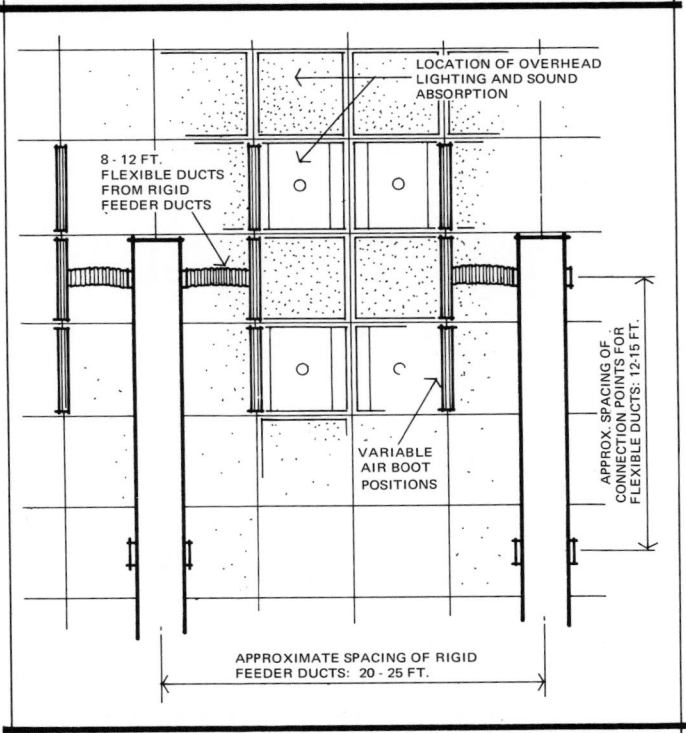

LOCATION OF OVERHEAD
LIGHTING AND SOUND
ABSORPTION

8 - 12 FT.
FLEXIBLE DUCTS
FROM RIGID
FEEDER DUCTS

APPROX. SPACING OF
CONNECTION POINTS FOR
FLEXIBLE DUCTS: 12-15 FT.

VARIABLE
AIR BOOT
POSITIONS

APPROXIMATE SPACING OF RIGID
FEEDER DUCTS: 20 - 25 FT.

of each joint detail must be standardized within close tolerances. There is some latitude in the stringency of this requirement; but this factor seems to have become progressively more precise and limited with the advance from the somewhat flexible character of masonry joints to the close tolerances that are associated with many metal-to-metal joints, electrical contact joints, or fluid service joints.

When the joint is taken into account, then, the somewhat rigid modular dimension may not necessarily equal the outside dimension of the major elements or components. The modular dimension is determined by the assembled component-joint combination. Similar to the length of a wave-cycle, then, the module is the dimension from a certain point (for example, mid-joint) to the next repeated identical point (mid-joint again). (See Figure 3–1.19.) Tolerances must therefore be established for components and joints both separately and together.

FIGURE 3–1.16

PARTITION LAYOUT ALTERNAT!VES:
MATRIX COORDINATION WITH STRUCTURE

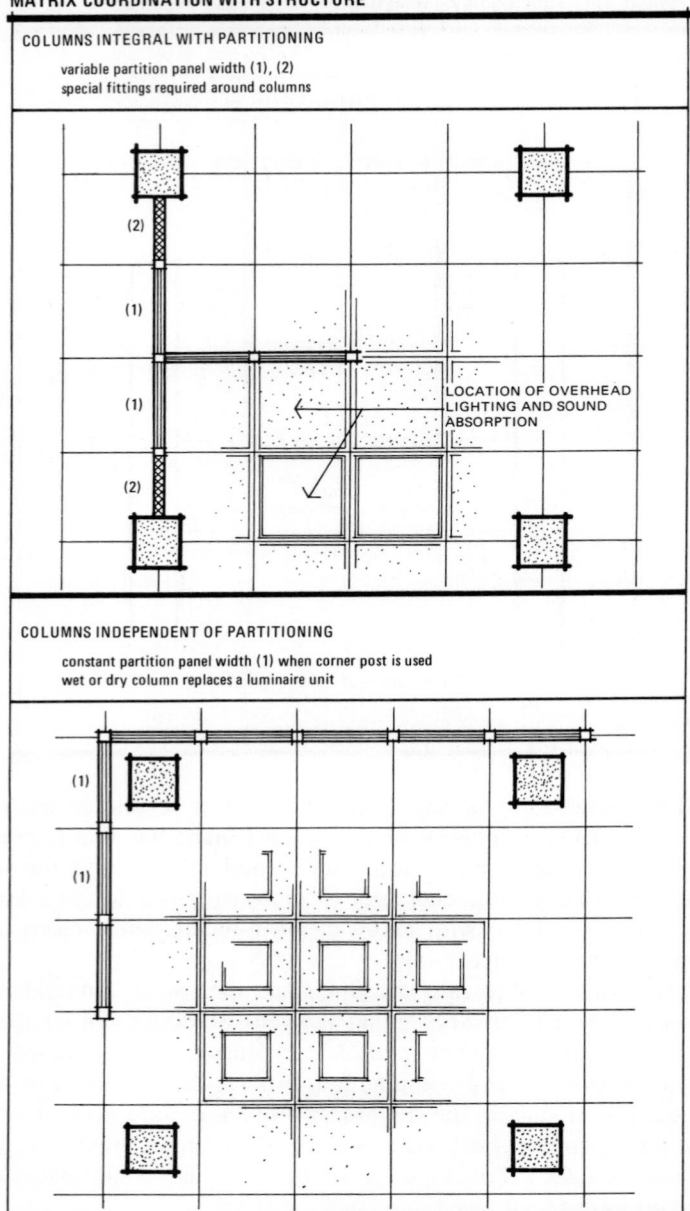

COLUMNS INTEGRAL WITH PARTITIONING

variable partition panel width (1), (2)
special fittings required around columns

(2)

(1)

(1)   LOCATION OF OVERHEAD
                                             LIGHTING AND SOUND
                                             ABSORPTION

(2)

COLUMNS INDEPENDENT OF PARTITIONING

constant partition panel width (1) when corner post is used
wet or dry column replaces a luminaire unit

(1)

(1)

FIGURE 3–1.17

PARTITION LAYOUT ALTERNATIVES:
MATRIX DEVELOPMENT INDEPENDENT OF STRUCTURE

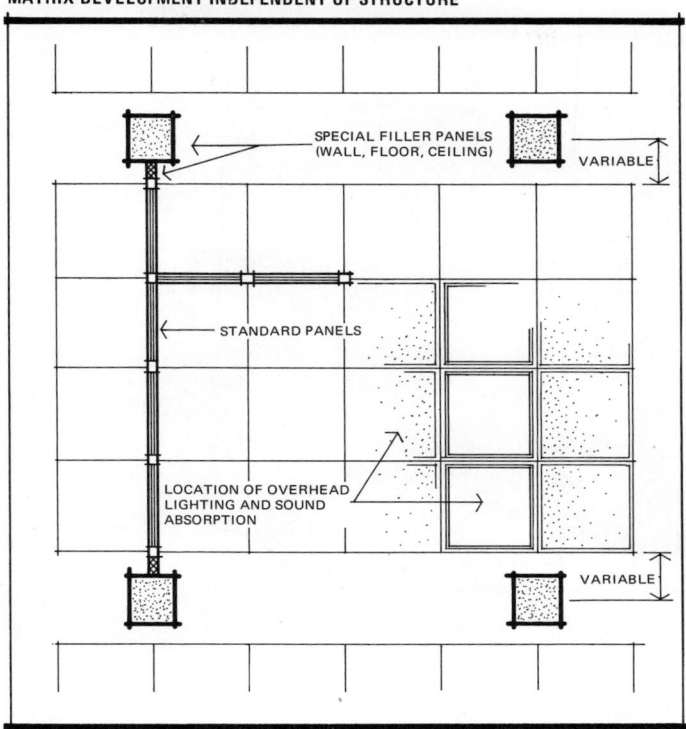

Labor Jurisdiction.   A related consideration in the design and installation of comprehensive environmental systems concerns disputes that may arise regarding the work to be done by various trades. As an example of attempts to clarify this situation, the following agreements generally apply to resolve disputes in the installation of ceiling elements. (Also see Figure 3–1.20.)

(1) The installation of typical acoustical ceilings is the work of the carpenter trades. The electrical trades will install all related wiring devices.

(2) When electric fixtures fit into a grid system, the grid systems that support both acoustical tile and electric fixtures are installed by the carpenter trades to the extent of providing parallel tees on two sides of the luminaires. If additional luminaire supports are required, these are installed by the electrical trades.

FIGURE 3–1.18A

MECHANICALLY COORDINATED MODULAR SYSTEMS

FIGURE 3–1.18B

FIGURE 3–1.18C

MECHANICALLY COORDINATED MODULAR SYSTEMS

The electricians will install all tees and hangers that support only luminaires. They will also mechanically secure and ground all fixtures.

(3) When the luminaires are suspended from a grid ceiling, the electrical trades will install all supporting hangers, install the luminaires, and install all supporting members. The carpenter trades will install all acoustical materials.

(4) For a luminous ceiling, the lamp channels on the upper ceiling are installed by the electrical trades. Beyond this, the general agreement is that if the diffuser panels and supporting grid are suspended from the electric lamp channels, the suspension system is installed by the electrical trades. If the diffuser panels and supporting grid are suspended independently of the electric fixtures, the suspension system is installed by the carpenter trades.

The controlling trade will then install the diffuser panels themselves. However, if an acoustical border or other acoustical panels are involved, these panels and related suspension elements will be installed by the carpenter trades.

FIGURE 3–1.19

THE PLANNING MATRIX:
JOINTS AND COMPONENT RELATIONSHIPS

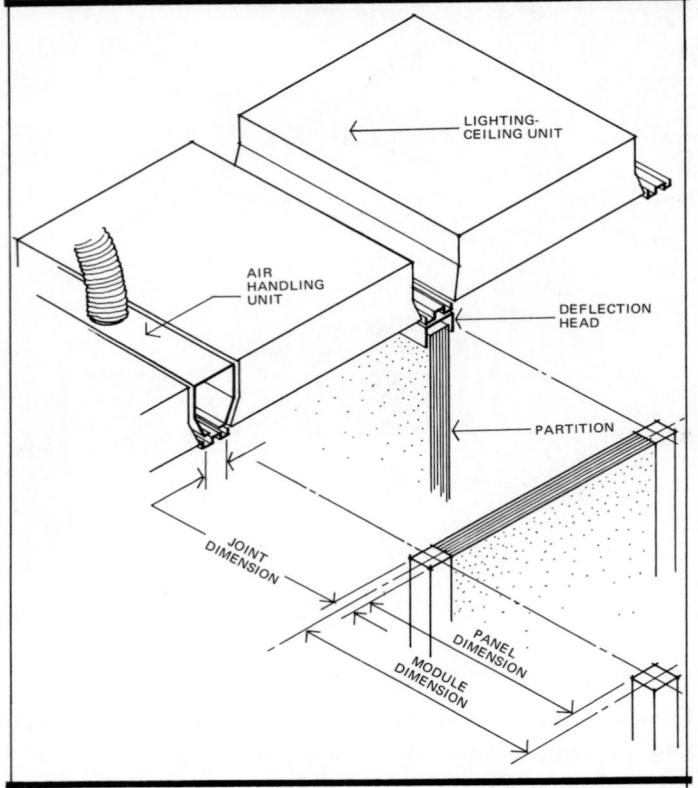

Flexibility   In order to maintain the usefulness of the building, the building system should provide sufficient flexibility to facilitate evolutionary changes in space utilization and layout. This is not to suggest that systems should be inordinantly over-sized to anticipate future change. Rather it implies the need for space allowances (i.e., space zoning) to facilitate and guide physical expansion of environmental system capacity and services in the future (see Figure 3–1.14). It also implies the need for suitable accessibility into these service spaces.

When this approach is effectively developed in the initial design, the interior space arrangement can be physically organized and reorganized as necessary without the need for fundamental alterations in the service distribution network.

FIGURE 3–1.20

TYPICAL ON-SITE JURISDICTION FOR INSTALLATION OF
MULTI-FUNCTION ASSEMBLIES

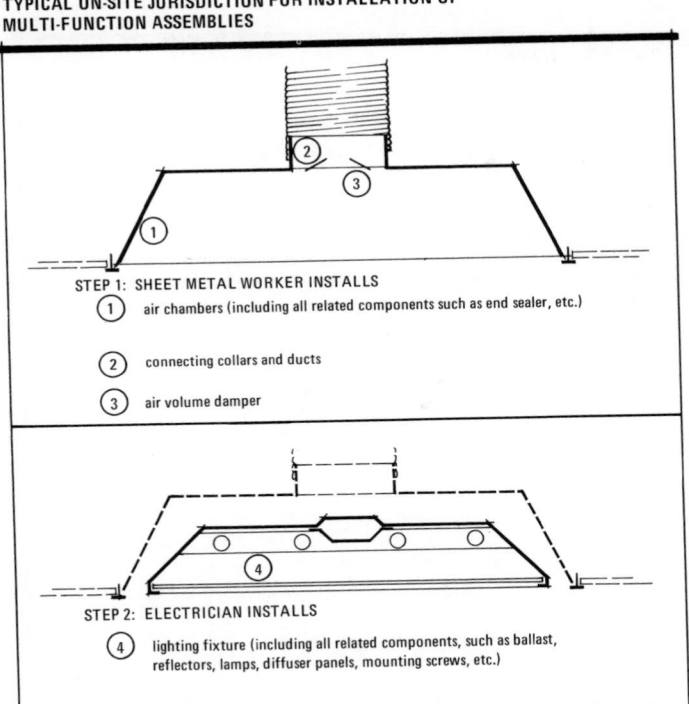

STEP 1: SHEET METAL WORKER INSTALLS

①   air chambers (including all related components such as end sealer, etc.)

②   connecting collars and ducts

③   air volume damper

STEP 2: ELECTRICIAN INSTALLS

④   lighting fixture (including all related components, such as ballast,
      reflectors, lamps, diffuser panels, mounting screws, etc.)

However, the term *flexibility* seems too broad for system design purposes. There are several categories or modes of change that can be implied
by this term.

Spatial Variety.   This is the system capability that permits the designer
to vary the permanent environmental characteristics of individual rooms.
It is mandatory that room-to-room variations be possible without violating
the basic structural, material, and dimensional discipline of the system.

Variations of this type most often respond to the specialized spatial
demands of individual but related activities (examples: distinctive system
variations for corridors, reception rooms, private offices, and general offices; or distinctive system variations for general classrooms, music rooms,
and laboratories).

In this sense, there should be a variety of component and finish alternatives for selection, without varying from the basic dimensional and
design discipline that has been established (see Figure 3–1.21).

FIGURE 3–1.21

THE PLANNING MATRIX:
REPRESENTATIVE CEILING VARIATIONS

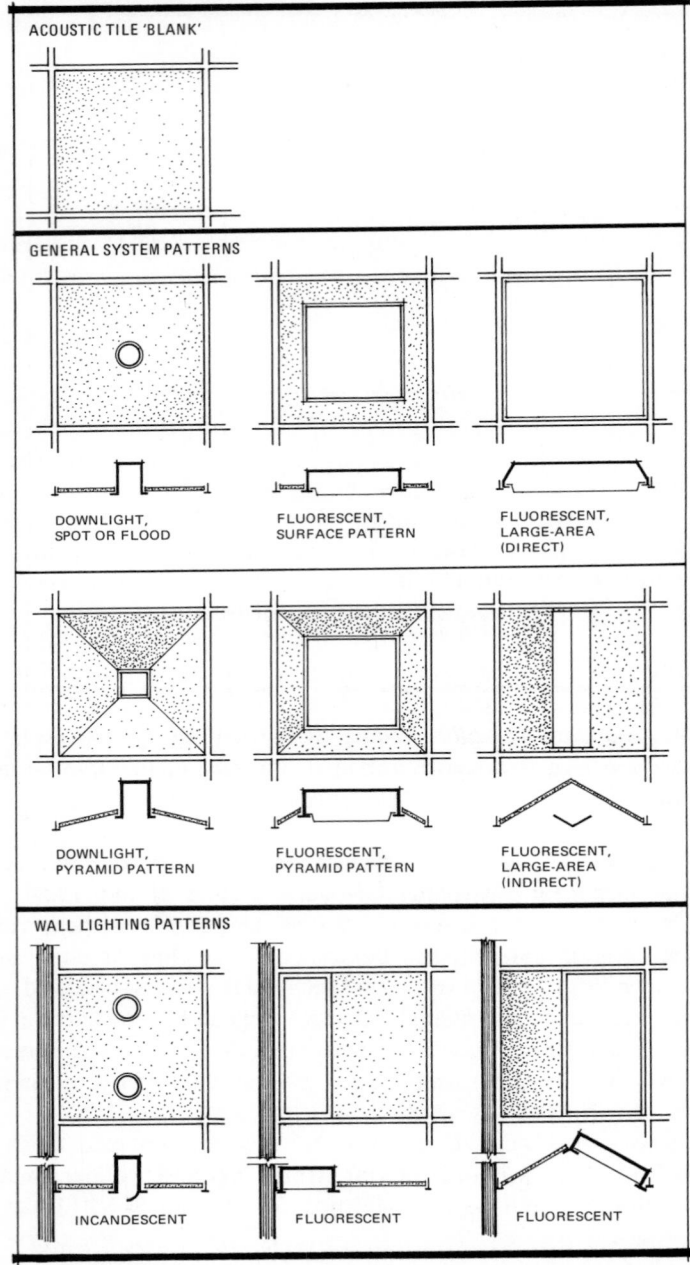

Short-Term or Intermittent Changes.   This category refers to situations that require conversion of space use with minimum time and effort. This may involve temporary expansion or subdivision of a space, which implies some portability of partitions and environmental devices.

Short-term variations may also involve changes in environmental conditions only; such as the changes in light characteristics that are required for an effective audio-visual space, or the changes in thermal requirements associated with lecture rooms and other spaces that have variable occupancy conditions. In these cases, the system response should involve switches, dimmers, and thermostatic control elements that will permit automatic or human-activated response to the changing environmental needs.

Long-Term or Permanent Changes.   It is likely that some rearrangement of facilities and layout will take place during the life of the building. The system should facilitate such changes by permitting the necessary relocation of partitions, ducts, supply and return air outlets, thermostatic and dimmer controls, acoustical and lighting devices, electric conduit and outlets, and control switches with a minimum of difficulty, disruption, and expense. This is not to infer that the system should be over-designed; but rather that it should be designed in a manner that will permit localized change with minimum labor, minimum material waste, and minimum disruption of adjacent areas. (See Figures 3–1.14 and 3–1.21.)

Expansion.   Expansion procedures should be defined in the initial design. Where relevant, process equipment should be placed in a position that will facilitate and accommodate an orderly and logical phasing of expansion plans. In this sense, anticipated future changes should require a minimum of demolition and disruptive interruptions within the existing building.

Bibliography

BOOKS

Flynn/Mills, *Architectural Lighting Graphics,* Reinhold, New York, 1962.

I.E.S. *Lighting Handbook,* 4th ed., Illuminating Engineering Society, New York, 1966.

Hopkinson, *Architectural Physics: Lighting,* Her Majesty's Stationary Office, London, 1963.

Phillips, *Lighting in Architectural Design,* McGraw-Hill, New York, 1963.

Noise Control in Buildings, Building Research Institute, no. 706, Washington D.C., 1959.

Beranek, *Acoustics,* McGraw-Hill, New York, 1954.

Handbook of Noise Measurement, 5th ed., General Radio Company, West Concord, Massachusetts, 1963.

Threlkeld, *Thermal Environmental Engineering,* Prentice-Hall, 1962.

Ashrae Handbook of Fundamentals,

Ashrae Guide and Data Book, American Society of Heating, Refrigerating, and Air Conditioning Engineers, New York, 1967.

Olgyay, *Design with Climate,* Princeton University Press, 1963.

Danz, *Sun Protection,* Frederick A. Praeger Co., New York, 1967.

SER-1: *Environmental Abstracts,*

SER-2: *Environmental Evaluations,*

SER-3: *Environmental Analysis,* University of Michigan Press, 1965.

Wachsmann, *The Turning Point of Building,* Reinhold, New York, 1961.

S.C.S.D.: *The Project and the Schools, E.F.L., Inc.,* Ford Foundation, 1967.

PERIODICAL REFERENCES

Architectural Record
Progressive Architecture
Architectural and Engineering News
Building Research (*Journal of the Building Research Institute*)
Illuminating Engineering (*Journal of the Illuminating Engineering Society*)
AIA Journal (*Journal of the American Institute of Architects*)
Building Construction

Index

Index

299